Science and Technology

London: H M S O

Researched and written by Publishing Services, Central Office of Information.

This publication is an expanded and updated version of the chapter on Science and Technology which appears in *Britain 1995: An Official Handbook*.

First published 1995

ISBN 0 11 701947 X

HMSO publications are available from:

HMSO Publications Centre
(Mail, fax and telephone orders only)
PO Box 276, London SW8 5DT
Telephone orders 0171 873 9090
General enquiries 0171 873 0011
(queuing system in operation for both numbers)
Fax orders 0171 873 8200

HMSO Bookshops
49 High Holborn, London WC1V 6HB
(counter service only)
0171 873 0011 Fax 0171 831 1326
68–69 Bull Street, Birmingham B4 6AD
0121 236 9696 Fax 0121 236 9699
33 Wine Street, Bristol BS1 2BQ
0117 926 4306 Fax 0117 929 4515
9-21 Princess Street, Manchester M60 8AS
0161 834 7201 Fax 0161 833 0634
16 Arthur Street, Belfast BT1 4GD
01232 238451 Fax 01232 235401
71 Lothian Road, Edinburgh EH3 9AZ
0131 228 4181 Fax 0131 229 2734
The HMSO Oriel Bookshop
The Friary, Cardiff CF1 4AA
01222 395548 Fax 01222 384347

HMSO's Accredited Agents
(see Yellow Pages)

and through good booksellers

Contents

Acknowledgments

The Central Office of Information would like to thank the following organisations for their co-operation in compiling this book: the Office of Science and Technology, the Department of Trade and Industry, the Ministry of Defence, the Ministry of Agriculture, Fisheries and Food, the Department of the Environment, the Department of Health, the Overseas Development Administration, the Foreign and Commonwealth Office, the British National Space Centre, the Engineering and Physical Sciences Research Council, the Particle Physics and Astronomy Research Council, the Biotechnology and Biological Sciences Research Council, the Medical Research Council, the Natural Environment Research Council, the Economic and Social Research Council, the Royal Society and other scientific organisations.

Cover Photograph Credit
David Kampfner.

Introduction

Britain[1] has a long tradition of research and innovation in science, technology and engineering in universities, research institutes and industry. Its record of achievement is in many ways unsurpassed, from the contributions of Isaac Newton to physics and astronomy in the 17th century (theory of gravitation and three laws of motion) to the work of Charles Darwin on the theory of evolution and the inventions of Michael Faraday in the 19th century (the first electric motor, generator and transformer). Great medical advances included the development of immunisation by Edward Jenner in the 18th century and the founding of antiseptic surgery by Lord Lister in the 19th century.

British achievements in science and technology in the 20th century have continued to be very significant. For example, fundamental contributions to modern molecular genetics were made through the discovery of the three-dimensional molecular structure of DNA (deoxyribonucleic acid) by Francis Crick, Maurice Wilkins, James Watson of the United States and Rosalind Franklin in 1953 and of cholesterol, vitamin D, penicillin and insulin by Dorothy Hodgkin.

Notable contributions in other areas over the past 25 years have been made by Stephen Hawking in improving the understanding of the nature and origin of the universe; Brian Josephson in superconductivity (abnormally high electrical conductivity at

[1] The term 'Britain' is used in this book to mean the United Kingdom of Great Britain and Northern Ireland. 'Great Britain' comprises England, Wales and Scotland.

low temperatures); Martin Ryle and Antony Hewish in radio astrophysics; and Godfrey Hounsfield in computer assisted tomography (a form of radiography) for medical diagnosis.

Much pioneering work was done during the 1980s. For example, in 1985 British Antarctic Survey scientists discovered the hole in the ozone layer over the Antarctic. In the same year Alec Jeffreys invented DNA fingerprinting, a forensic technique which can identify an individual from a small tissue sample. More recently there have been several British breakthroughs in genetics research, including the identification of the gene in the Y chromosome responsible for determining sex, and the identification of other genes linked to diseases, including cystic fibrosis and a type of inherited heart disease. A vaccine has been developed to protect against cancer associated with the Epstein Barr virus. Gene therapy has begun on the treatment of cystic fibrosis. The world's first pig to have a genetically modified heart has been bred by scientists at Cambridge University, an important milestone in breeding animals as organ donors for people.

Nobel Prizes for Science have been won by over 70 British citizens, more than any other country except the United States. The two most recent (1993) prizewinners were Richard Roberts (medicine) and Michael Smith (chemistry).

International collaboration has long been a tradition in science and is increasingly important today. Many British research teams are engaged in major programmes in partnership with their overseas counterparts. However, while teams of scientists dominate present-day discoveries more than the pioneering individuals of the past, British academics and inventors continue to establish worldwide reputations.

Science, engineering and technology are fundamental to the economic success of any internationally competitive country. The research carried out in industry and by Britain's publicly funded higher education institutions and research institutes forms a vital contribution to the health of the economy and the country's future industrial development.

The Government considers that public funding should primarily support the pursuit of basic scientific knowledge, while industry should bear the main responsibility for the commercial development of scientific advances. Its White Paper *Realising Our Potential: A Strategy for Science, Engineering and Technology* (see Further Reading, p. 105) outlining future policy on science, engineering and technology, published in 1993, stresses the need for closer links between the science base and industry.

This book outlines the recent major policy changes in science and technology, and describes the new organisation at the beginning of 1995. It also outlines some of the major research projects. An Appendix gives a selection of major scientific achievements in Britain.

Research and Development Expenditure

Total expenditure in Britain on scientific research and development (R & D) in 1993 was £13,752 million (see Table 1), 2.2 per cent of gross domestic product (GDP), similar to the average in the European Union (EU). Just over half was funded by industry and nearly a third by government. Significant contributions were also made by private endowments, trusts and charities. About 279,000 people (full-time equivalent staff) were employed on R & D in 1993: 164,000 in business, 66,000 in higher education, 35,000 in government and 15,000 in private non-profit-making bodies.

Many industries fund their own research and run their own laboratories. Large companies like ICI, British Telecom (BT) and BP develop new types of products, equipment and processes. Industry also funds university research and finances contract research at government establishments and individual research organisations and at contract research organisations. A large amount of funding for research, particularly in medicine, also comes from Britain's charitable organisations. Some charities have their own laboratories and offer grants for outside research. Contract research organisations carry out R & D for companies and are playing an increasingly important role in the transfer of technology to British industry.

Total spending on R & D undertaken in British businesses in 1993 was £9,126 million, about 1.5 per cent of GDP. This represented a rise of 19 per cent compared with 1989, but in real

Table 1: R & D Undertaken in Britain 1993

£ million

Sector providing the funds	Sector carrying out the work: Government	Higher education	Business enterprise	Private non-profit	Totals
Government	1,599	1,556	1,129	162	**4,446**
Higher education	4	100	0	0	**104**
Business enterprise	203	176	6,542	240	**7,161**
Private non-profit	51	283	0	96	**430**
Abroad	36	151	1,398	26	**1,611**
Total	**1,893**	**2,266**	**9,069**	**524**	**13,752**
of which:					
Civil	*1,238*	*2,230*	*7,710*	*513*	*11,692*
Defence	*654*	*36*	*1,359*	*11*	*2,060*

Source: Central Statistical Office.

terms there was a fall of 3 per cent.[2] Pharmaceuticals accounted for 18 per cent of expenditure: £1,651 million. About 77 per cent of expenditure was by manufacturing concerns and 23 per cent by services. Expenditure by the main industrial sectors is given in Table 2. A breakdown for expenditure between civil and defence categories is shown in Table 3. Chemical products and service businesses accounted for over half of civil R & D, whereas defence R & D is dominated by aerospace and electrical products. R & D expenditure on manufactured products represented 2.4 per cent of total sales. Pharmaceuticals had by far the largest proportion of

[2] Statistics in this paragraph and in Tables 2 and 3 are taken from the initial survey of business enterprise R&D produced by the Central Statistical Office (CSO) in December 1994. Revised figures from the survey have been incorporated in the CSO's statistics for gross domestic expenditure on R&D in 1993, issued in March 1995, which is the source for Table 1.

Table 2: Expenditure on R & D Undertaken in British Businesses by Broad Product Groups

£ million

	1986	1989	1990	1991	1992	1993
Manufacturing	4,604	5,901	6,490	6,273	6,583	6,996
of which:						
Chemicals	*1,039*	*1,692*	*2,027*	*2,003*	*2,262*	*2,495*
Mechanical engineering	*261*	*345*	*374*	*400*	*433*	*560*
Electrical machinery	*1,498*	*1,525*	*1,662*	*1,454*	*1,511*	*1,643*
Transport equipment	*419*	*548*	*590*	*609*	*640*	*681*
Aerospace	*830*	*1,090*	*1,124*	*1,125*	*1,027*	*863*
Other manufacturing	557	701	713	682	711	755
Services	1,347	1,749	1,828	1,861	1,907	2,130
Total	**5,951**	**7,650**	**8,318**	**8,135**	**8,489**	**9,126**

Source: Central Statistical Office.

Note: Differences between totals and the sums of their component parts are due to rounding.

Table 3: Expenditure on Civil and Defence R & D Undertaken in British Businesses by Broad Product Group 1993

£ million

	Civil	Defence	Total
Manufacturing	5,809	1,187	6,996
of which:			
Chemicals	*2,469*	*26*	*2,495*
Mechanical engineering	*396*	*164*	*560*
Electrical machinery	*1,268*	*375*	*1,643*
Transport equipment	*622*	*59*	*681*
Aerospace	*372*	*491*	*863*
Other manufacturing	*682*	*73*	*755*
Services	1,959	172	2,130
Total	**7,767**	**1,359**	**9,126**

Source: Central Statistical Office.

Note: Differences between totals and the sums of their component parts are due to rounding.

sales spent on R & D—21 per cent, followed by aerospace (9 per cent). R & D expenditure is heavily concentrated in the South East, which accounted for 56 per cent of expenditure.

The numbers employed on R & D in business concerns fell by 6 per cent between 1989 and 1993, to 164,000: 86,000 scientists and engineers, 40,000 technicians, laboratory assistants and draughtsmen, and 37,000 administrative and other support staff.

Of the ten companies with the largest annual expenditure on R & D (see Table 4), four—Glaxo, SmithKline Beecham, Zeneca (formerly part of ICI) and Wellcome[3]—are in pharmaceuticals.

[3] Glaxo made a successful takeover bid for Wellcome in 1995.

Table 4: Annual Company Expenditure on R & D

	R & D annual expenditure (*£ million*)	R & D as % of sales
Glaxo	739	15.0
SmithKline Beecham	575	9.3
Shell Transport and Trading	529	0.6
Unilever	518	1.9
Zeneca	490	11.0
GEC	398	7.1
Wellcome	326	16.0
Rolls-Royce	253	7.2
British Petroleum	237	0.7
BT	233	1.8

Source: DTI R & D Scoreboard.

Note: R & D expenditure includes expenditure overseas.

Government Role

The Government intends that future spending on science and technology should be more closely related to meeting Britain's economic and other needs and enhancing wealth creation. Science and technology issues are the responsibility of the President of the Board of Trade, who is supported by the Office of Science and Technology (OST)—part of the Department of Trade and Industry (DTI). It is headed by the Government's Chief Scientific Adviser. The Government seeks to strengthen the science base. Its policies on science, engineering and technology are being developed alongside policies on industry, taking account of the important role of research and development in improving economic performance and the quality of life.

In addition to its role in overall science policy and co-ordination, the OST is responsible for the Science Budget and the six government-financed research councils:

—the Engineering and Physical Sciences Research Council (EPSRC);

—the Particle Physics and Astronomy Research Council (PPARC);

—the Medical Research Council (MRC);

—the Natural Environment Research Council (NERC);

—the Biotechnology and Biological Sciences Research Council (BBSRC); and

—the Economic and Social Research Council (ESRC).

OST funding provides assistance for research, through the research councils, in the following ways:

—awarding grants and contracts to universities and other higher education establishments and to research units;

—funding research council establishments;

—supporting postgraduate study; and

—subscribing to international scientific organisations.

The OST also supports universities (whose main source of funding is through the appropriate higher education funding council) through programmes administered through the Royal Society and the Royal Academy of Engineering (see pp. 43–4).

Strategy

The White Paper *Realising Our Potential: A Strategy for Science, Engineering and Technology* contains the first major policy review on science for over 20 years. The Government's strategy, following the White Paper, is to improve Britain's competitiveness and quality of life by maintaining the excellence of science, engineering and technology in Britain, and by:

—developing stronger partnerships with and between the science and engineering communities, industry and the research councils;

—supporting the science and engineering base in order to advance knowledge, increase understanding and produce highly educated and trained scientists and technologists;

—contributing to international research efforts, particularly European research;

—promoting the public understanding of science, technology and engineering; and

—ensuring that government-funded research is conducted efficiently and effectively.

Changes in the Organisation of Science and Technology

A number of changes to the organisation and direction of government-funded science, engineering and technology have been made following the White Paper:

—A new Council for Science and Technology has been set up, chaired by the Chancellor of the Duchy of Lancaster. Its members include senior people from industry and academic life. It provides independent advice to government on strategic issues of science and technology, including expenditure priorities.

—A new *Forward Look*, first published in April 1994, is published by the Government each year. This gives industrialists and researchers a regular statement of the Government's strategy.

—A new Technology Foresight Programme, launched in February 1994, is intended to strengthen links between industry, academia and government. It aims to identify market opportunities for the British economy over the next 10–20 years and the strategic areas of science and technology necessary to exploit them, and to develop networks which will engage the business and scientific communities jointly in long-term foresight activity. The programme is being taken forward through 15 sector panels and a joint industry–academic steering group chaired by the Chief Scientific Adviser. Panel reports and an overall assessment from the steering group were published in 1995 (see Further Reading).

—The research council system was restructured with effect from 1 April 1994 (see p. 18). Management structures have been modified, and councils now have part-time chairmen selected to bring in industrial and commercial experience. The Director-General of Research Councils within the OST advises ministers on the performance and needs of the science and engineering base.

—The customer-contractor principle has been maintained and strengthened.

—The OST now plays the major role in co-ordinating science and technology issues involving more than one department.

—Special importance is now attached to co-ordination of European and other international negotiations.

Public Awareness

A major policy initiative in the White Paper was the announcement of a campaign to promote the public awareness of science, engineering and technology. By improving the climate for innovation and providing companies with access to technology and best practice, the Government hopes that industry will facilitate the change towards a better understanding and exploitation of the benefits of science, engineering and technology. The campaign has two aims:

—to change public perceptions of science and engineering so that there is greater appreciation of the contribution of science and engineering to Britain's economic and social development; and

—to increase public understanding of science and technology so that public debate of scientific and technological issues can be better informed.

The first major event of the campaign was the National Week of Science, Engineering and Technology, held in March 1994. About 1,200 events were held in 230 towns and cities across Britain. These were organised, with financial support from the Government, by the British Association for the Advancement of Science. A second week was held in March 1995 with around 3,000 events, and this is now intended to be an annual event. As part of the campaign, the Government also supports organisations and events such as the Edinburgh International Science Festival (see p. 48).

In 1995 the Government established a committee under Professor Sir Arnold Wolfendale to review the steps being taken to encourage scientists, engineers and research students to contribute to public understanding of science, engineering and technology with a view to recommending how this understanding can be improved.

LINK

The LINK scheme, administered by the OST with funding principally from the Department of Trade and Industry and the EPSRC, is designed to bridge the gap between industry and the research base for the benefit of the British economy. It encourages collaboration on high quality, commercially relevant research between industry and higher education institutions and other research base organisations. Under the scheme, government departments and research councils fund up to 50 per cent of the cost of research projects, with industry providing the remainder. The participating industrial partners in LINK projects are responsible for the commercial exploitation of research outcomes. By March 1995, 570 projects, worth over £300 million, had been started, involving over 800 companies and 130 scientific institutions. Many of these projects have led to new products, processes and services being marketed by British companies.

In March 1995 the Government announced that the scheme would be modified to make it more effective. A new board will oversee the scheme, and future LINK programmes will be responsive to priorities identified by the Technology Foresight Programme.

Science Budget

Total net government R & D expenditure (both civil and defence) in 1993–94 was £5,387 million, of which £3,088 million was devoted to civil science. The Science Budget for 1995–96 totals £1,281.7 million (see Table 5), compared with £1,236.5 million for 1994–95. Government funding through the Science Budget has increased by over 20 per cent in real terms since 1984–85.

About 5 per cent of the Budget is allocated to priority initiatives designed to advance the Government's science and engineering technology policy as set out in the 1993 White Paper. The initiatives fall into three main areas:

—improved interaction with industry and commerce;

—enhancements to basic and strategic science; and

—enhancements to 'people-related' programmes.

Improved Interaction with Business

Among the schemes in this area are:

—The Realising Our Potential Awards (ROPA) Scheme, which was announced in February 1994 and has the dual aims of enhancing collaboration between the science and engineering

Table 5: Science Budget 1995-96

	£ million	%	Priority initiatives included in 1995-96 allocations *(£ million)*
EPSRC	365.7[a]	28.5	21.1
MRC	277.8	21.7	14.0
PPARC	196.4	15.3	8.3[b]
BBSRC	161.6	12.6	12.2
NERC	155.5	12.1	3.2
ESRC	61.2	4.8	1.3
CCL[c]	1.5	0.1	1.5
Royal Society	20.8	1.6	1.5
Royal Academy of Engineering	2.6	0.2	0.8
OST initiatives and other payments	38.5	3.0	3.3
Total	1,281.7	100	67.2

Source: Office of Science and Technology.

[a] Includes £7.2 million on high-performance computing on behalf of all the research councils.

[b] Includes international subscriptions reserve of £8 million.

[c] The new Council for the Central Laboratory of the Research Councils (see p. 26).

base and industry and of providing to researchers already working with industry funds to carry out research in an area of their own choice. A pilot scheme involved three research councils—the EPSRC, the BBSRC and the MRC—and about 240 projects were funded by the councils. This scheme is being extended to all research councils and the budget for 1995–96 is £21.6 million.

—The industry-led Co-operative Awards in Science and Engineering (CASE) scheme, launched by the research councils, which is supporting over 120 research students.

—LINK (see p. 13) where an additional £3 million has been earmarked for research councils to contribute to new LINK programmes.

—The Innovative Manufacturing Initiative, launched in 1994, which offers joint funding, provided equally by the research councils and industry, and is intended to cover innovation in products, in the manufacturing process and in the business processes through which they are marketed.

Enhancement of Underpinning Strategic Science

Enhanced support for chemistry has been provided within the BBSRC, EPSRC and NERC, and it is now proposed to extend this into mainstream physics, mathematics and medicine. In addition, strategic areas have been identified where additional support could produce benefits for Britain through emerging technological and market opportunities:

—increasing the understanding of genetic composition;

—improved understanding of the immune response and its role in the control of infectious diseases which is necessary to support developments in vaccines—a new independent Edward Jenner Institute for Vaccine Research, jointly funded by Glaxo, the MRC, the BBSRC and the Department of Health, is being built at the BBSRC's Institute of Animal Health at Compton (Berkshire);

—research on various aspects of bioprocessing innovation;

—additional work on research connected with products from plants;

—environmental diagnostics; and

—'cognitive engineering', covering communication between people and computers.

Development of People

Additional funds were provided in 1994–95 to the Royal Society and the Royal Academy of Engineering, which were mainly targeted at developing the individual. Further funds are being provided for the development of these schemes. The Royal Society will be increasing its research fellowships to 255 and will also be launching on a pilot basis a new fellowship scheme for younger scientists. The Royal Academy of Engineering will be launching a programme of industrial secondments for academic engineers, building on a pilot scheme.

Government Departments

Among government departments, the Ministry of Defence (MoD—see p. 33) has the largest research budget. The main civil departments involved are the Department of Trade and Industry (see p. 28), the Ministry of Agriculture, Fisheries and Food (MAFF—see p. 34) and the Department of the Environment (see p. 35).

Research Councils

Each research council is an autonomous body established under Royal Charter, with members of its governing council drawn from the universities, professions, industry and government. Councils conduct research through their own establishments and by supporting selected research and postgraduate training in universities and other higher education institutions.

In April 1994 two new councils—the EPSRC and the PPARC—were created, dividing between them most of the portfolio of the former Science and Engineering Research Council (SERC). A third new council, the Biotechnology and Biological Sciences Research Council, took over responsibility for the programme of the former Agricultural and Food Research Council and for the work of the SERC in biology and biotechnology.

In addition to funding from the OST, the research councils receive income for research commissioned by government departments and from the private sector. Income from commissioned research is particularly important for the BBSRC and NERC: receipts of these councils in 1994–95 were expected to offset over 20 per cent of gross expenditure.

Engineering and Physical Sciences Research Council

The EPSRC, the largest research council, supports underpinning research in engineering and the physical sciences for a very wide range of industries including electronics, chemicals, pharmaceuticals, transport, information technology, communications, construction, manufacturing, energy, aerospace, food, hotels and catering, banking, finance, leisure, and retailing. About 80 per cent of its research and training investment is spent in higher education institutions.

Priority areas in which the EPSRC is working are:

—innovative manufacturing and clean technology, underpinned by research in a range of generic technologies—information technology, materials, design, systems and production engineering and processing technology;

—research and training in core engineering and physical science disciplines to maintain their health and to aid the EPSRC's initiatives and key technologies; and

—continuing access to high-quality national and international facilities for British researchers.

Particle Physics and Astronomy Research Council

The major task of the PPARC is to ensure continuity of the country's long tradition of excellence in research into fundamental physical processes through studies in particle physics, astronomy and astrophysics, and solar system science. The PPARC maintains four research establishments under common management: the Royal Greenwich Observatory at Cambridge, the Royal Observatory at Edinburgh and two overseas observatories on La Palma in the Canary Islands and on Hawaii (see p. 55). The PPARC is a major source of funding for many leading university physics departments in Britain.

Specific research projects to be undertaken over the next decade include:

—physics of the 'Big Bang': the origin of mass and the nature of the fundamental particles and forces;

—symmetry of matter and anti-matter;

—structure and evolution of the universe since the 'Big Bang';

—the resolution of the 'missing mass' or 'dark matter' problem: the difference between the visible mass of the universe and that required to explain its physical properties;

—star formation and evolution;

—coupling of the Sun's emissions to the Earth's ionosphere and atmosphere; and

—the formation and evolution of the solar system.

The Council's work is in fields where international co-operation is particularly important; substantial contributions are made to the European Space Agency (ESA) and the European Laboratory for Particle Physics (CERN, see p. 51) where the proposed Large Hadron Collider is a major research interest.

Through the PPARC, Britain is taking part in one of the most important ground-based astronomical projects of the 1990s—Gemini, an international collaboration to build two 8-metre telescopes that will give all-sky coverage from sites in Hawaii and Chile. Britain is participating in the design, construction and instrumentation of the telescopes.

Biotechnology and Biological Sciences Research Council

The BBSRC sustains a broad base of interdisciplinary research and training to underpin the biology-based industries—agriculture, pharmaceuticals, chemicals and food.

The Council operates underlying science-led research programmes organised in areas reflecting the different levels of biological structure and research programmes to develop further opportunities arising from such research. Science-led programmes include work on biomolecular science; genes; cell biology and biochemistry; plant, animal and microbial science; systems and engineering. Directed programmes include work on agricultural systems and on chemicals and pharmaceuticals (including the design, development and efficient manufacture of bioproducts) and research of relevance to the food industry. Research is carried out at eight institutes, at Horticulture Research International and in higher education institutions.

The BBSRC has an important role in the training of scientists and engineers: it has over 2,000 research studentships and 75 post-doctoral fellowships. Use is made of the two main mechanisms for supporting postgraduate training with an industrial orientation: CASE awards and awards under the Teaching Company Scheme (see p. 33).

Recombinant DNA technologies have helped to produce novel domestic animal vaccines against economically important diseases and these have been successfully commercialised. Advances in gene technologies have led to the development of fruit with improved storage capacity and hence longer shelf life in supermarkets. It is also possible to produce human pharmaceutical proteins such as blood clotting factors in animal milk, from which these proteins are easily extracted in a pure form. The use of biotransformations has led to the biosynthesis of industrially important pure chemicals which are key building blocks in some anti-viral agents.

Medical Research Council

The MRC is the main government agency supporting medical research. It promotes and supports research and training aimed at maintaining and improving human health, advancing knowledge and technology, and providing well-trained staff to meet the needs of user communities, including the providers of health care and the biotechnology, food, health care, medical instrumentation, pharmaceutical and other biomedical-related industries.

The MRC is unusual among the research councils in that about half its research expenditure is spent in its own institutes and units, which have close links with universities and medical schools, the rest being spent mainly in higher education institutions. The Council has three large institutes—the National Institute for Medical Research at Mill Hill in London, the Laboratory of Molecular Biology in Cambridge and the Clinical Sciences Centre (opened in 1993) at Hammersmith Hospital, London—and 40 research units and a number of smaller teams.

Areas in which the MRC is conducting research include:

—Molecules and cells, and inheritance and development—work in these areas is contributing to the identification of novel cellular targets for new pharmaceutical agents and to the detection, prevention and treatment of diseases with a genetic component.

—Infections and immunity—these areas include work on developing new vaccines and identifying novel targets for antiviral chemotherapy. The MRC has a major research programme on HIV infection and AIDS.

—Environment—the Council's strategy is to determine the health impact of particular environmental hazards; the MRC has established a research centre in Mechanisms of Human Toxicity and, with the Departments of the Environment and Health, set up an Institute for Environment and Health.

—Cancer—the Council's activity is aimed at increasing basic understanding of tumour biology, developing methods of detecting cancer, identifying preventive strategies and developing and evaluating better methods of early detection and treatment. The MRC's work complements that of the medical research charities and involves important international collaboration.

—Neurosciences and mental health—this is an area to which the Council commits very considerable expenditure in order to extend understanding of the central nervous system, explore diseases of the nervous system and develop new methods of diagnosis, prevention and treatment. Support is given to research on psychoses (including schizophrenia), neuroses, dementia, addictions, child psychiatry and neurological disorders.

—Systems-oriented research—this concerns main body systems; particular areas are cardiovascular studies, haematology, respiratory disease, reproduction, calcium metabolism and bone disorders, rheumatology, dental research, nutrition and metabolism.

—Health services and public health research—the Council's work is focused on implementing medical knowledge to improve health and address needs for which effective medical interventions exist or could soon be developed. The MRC's work complements that of the Government's Health Departments, the National Health Service (NHS) and the ESRC.

Scientists working at the MRC Human Genetics Unit in Edinburgh have succeeded in creating a strain of laboratory mice carrying the defective gene that causes cystic fibrosis. The Edinburgh mouse, unlike other models, closely mimics the early stages of lung disease in patients. The cystic fibrosis mouse has already proved valuable in developing and testing the safety and effectiveness of new drug and gene therapies before conducting clinical tests on people. These trials are being undertaken at the Royal Brompton Hospital, London (see p. 70).

Natural Environment Research Council

The NERC has an R & D provision of £155 million for 1995–96 and its wide-ranging activities fall into the following categories:

—Earth observation—advanced research methodology for measuring the properties of the Earth's atmosphere, land, sea and the surface. Since April 1994 the NERC's responsibilities cover support for the whole area of Earth observation, including instrument development;

—Earth sciences—structure, composition, processes and mineral resources of the Earth (geology, geophysics, geochemistry, palaeontology and aspects of physical geography);

—marine sciences—physical, chemical and biological characteristics and processes of the seas and sea floor;

—atmospheric sciences—physical and chemical processes determining weather and climate, atmospheric pollution and air-surface interactions;

—terrestrial and freshwater sciences—physical, chemical and biological characteristics and processes of land and freshwater; and

—polar sciences—physical, chemical and biological aspects of the Antarctic and Arctic environments (atmospheric, cryospheric, marine, earth and life sciences).

The Council supports research in 15 NERC institutes, units and research centres as well as research and training in universities. It also provides a range of facilities for use by the environmental science community including a research fleet. NERC institutes include the British Geological Survey, the Institute of Terrestrial Ecology, the Institute of Oceanographic Sciences Deacon Laboratory and the British Antarctic Survey.

A new national centre for oceanographic science and technology, the Southampton Oceanography Centre, will be officially opened in 1995. It is intended to be a world-class centre of excellence in this area and to be a national focus for research, training and support in oceanography, geology and aspects of marine technology and engineering.

The NERC has a substantial income from commissioned research and other services making use of its expertise. Programmes such as those of the British Geological Survey are crucial for environmental conservation; the Land Ocean Interaction Study (see p. 83) is also important as are programmes in marine, atmospheric and terrestrial and freshwater sciences and in Antarctic research.

Economic and Social Research Council

The ESRC supports research to increase understanding of social and economic change in order to enhance industrial competitiveness and the quality of life and to contribute to the effectiveness of public services and policy. All research funded by the ESRC is conducted in higher education institutions or independent research institutes.

The Council devotes a substantial part of its expenditure to high quality training in research to aid the development and maintenance of a first-class research capacity in the social sciences. Over 1,000 postgraduate students are supported by the Council through research training awards, advanced course awards and fellowships. Through its participation in the DTI's Teaching Company Scheme, the ESRC promotes partnerships between social scientists and business by placing graduates in companies for short-term projects.

The main areas for the Council's research are economic performance, environmental change, social cohesion, government and public services, health and welfare, and human development and learning. The priority for the allocation of funds in the next three years will be work on the study and enhancement of innovation in the British economy. The ESRC is a joint sponsor of an initiative, being managed by the EPSRC, on innovation in manufacturing.

Council for the Central Laboratory of the Research Councils

The Council for the Central Laboratory of the Research Councils (CCL) was established as an independent body with effect from 1 April 1995. It is responsible for the Daresbury Laboratory, in Warrington, Cheshire, and the Rutherford Appleton Laboratory, in Chilton, Oxfordshire, which formed the constituent parts of the Daresbury and Rutherford Appleton Laboratories (DRAL). The income of DRAL in 1994–95 was about £93 million and about

2,000 staff are employed. Following a review, the Government decided that DRAL should remain in public ownership, as an independent body, with strong emphasis on operating in a commercial manner; its customers are expected to continue to include five of the six research councils.

The laboratories provide experimental facilities too large and complex to be housed by individual academic institutions, and also undertake contract research. For example, Chilton has one of the most powerful lasers devoted to civil research in Europe—Vulcan—which is capable of simulating conditions inside stars. The ISIS pulsed neutron source is used by scientific teams from over 20 countries, including those working on research on high-temperature superconductors. In recent years the Synchrotron Radiation Source at Daresbury has enabled scientists to establish the structure of viruses that cause foot and mouth disease and sleeping sickness. An important new ultraviolet microscope has also been installed at Daresbury (see p. 61).

Government Departments

The main role of government departments is to stimulate innovation in industry so as to enhance industrial competitiveness in Britain and overseas.

Department of Trade and Industry

Direct government support for research in industry is led by the Department of Trade and Industry. In 1994–95 it spent an estimated £299 million on R & D covering general industrial innovation, aeronautics, space (see p. 37) and support for statutory, regulatory and policy responsibilities. A further £53 million was spent on energy R & D programmes. Some £22 million was spent on nuclear R & D, mainly on nuclear fusion. Non-nuclear R & D includes research in support of renewable and novel sources of energy, offshore oil and gas technology, and clean coal technologies.

DTI Industrial Innovation Programmes

The DTI's policy on science and technology builds on the commitment contained in the 1993 White Paper to improve the partnership between government, industry and the science base. Additional measures were set out in a 1994 White Paper on competitiveness.[4] The balance of the DTI's industrial innovation programmes has shifted away from supporting technology

[4] *Competitiveness: Helping Business to Win*. See Further Reading.

generation towards concentrating on the exploitation and transfer of technology and the promotion of innovation. It is improving companies' access to local innovation services through the network of around 200 business advice centres—Business Links—which are being set up to bring together in a single point of access organisations supporting enterprise, such as local companies, Training and Enterprise Councils, chambers of commerce, local authorities and enterprise agencies. They offer a full range of services—including innovation and technology transfer—to established medium-sized and small firms.

SUPERNET, a new network of centres of technological excellence, was launched in November 1994 to assist innovation and technology counsellors in Business Links to deliver technological services to clients more easily. Business Links are using SUPERNET to ensure that smaller firms obtain help to boost their competitiveness by drawing on the skills and expertise of universities and research and technology organisations.

In addition, the DTI is:

—encouraging industry to collaborate with the science base in R & D projects under the LINK initiative (see p. 13);

—putting more effort into helping firms of all sizes work together to undertake R & D projects, including those under the EUREKA initiative (see p. 50);

—facilitating companies' access to technology from overseas and helping small and medium-sized concerns to identify technological opportunities and potential partners, both in Britain and overseas; and

—concentrating support from the DTI's innovation budget for industry R & D collaboration on those projects which will result in exceptional economic benefits.

The DTI is also continuing direct single-company support to smaller firms under the SMART competition and the SPUR scheme for the development of new products and processes. SPUR (Support for Products Under Research) offers grants to businesses with up to 250 employees for new product and process development demonstrating a significant technological advance. SPUR grants of £46 million have helped over 500 projects since the scheme began in 1991. SMART (the Small Firms Merit Award for Research and Technology) is an annual competition providing grants to individuals and businesses with fewer than 50 employees in support of innovative technological projects with commercial potential. Over 1,200 projects have been assisted since the scheme began in 1986.

To carry out these measures, the DTI's budget allocation for support for industrial innovation is £108.8 million in 1995–96.

Aeronautics

The DTI's Civil Aircraft Research and Demonstration Programme (CARAD) supports research and technology demonstration in the aircraft and aeroengine industry, helping it to compete effectively in world markets. The programme is part of a national aeronautics research effort, with over half of the research work supported being conducted in industry and the universities, and the remainder at the Defence Evaluation and Research Agency (see p. 34). Priority areas are aircraft exhaust emissions, advanced materials, and safety problems such as explosion hazards. In 1995–96 the provision for civil aircraft research and demonstration is £22.6 million.

Launch Aid is a means of providing government assistance for specific development projects in the aerospace industry. The civil version of the EH101 British/Italian multi-role helicopter is

receiving Launch Aid assistance, and a significant offer of Launch Aid has been made to Short Brothers for the Learjet 45 project for 1995–96 and 1996–97, while other Launch Aid projects are under consideration.

Industrial Research Establishments

The DTI has three research establishments. Their primary role is to provide the Government with an effective source of scientific and technological expertise. They supply technological services to industry, undertake research commissioned by industry, and are involved in a variety of international activities.

—The Laboratory of the Government Chemist (LGC) is the focus within government for chemical measurement. It provides both public and private sectors with a comprehensive service based on analytical chemistry and promotes sound chemical measurement in Britain in support of the National Measurement System.

—The National Engineering Laboratory (NEL) carries out a range of technical services, including R & D, testing and consultancies in engineering and related disciplines. It also maintains British Standards of flow measurement, which are of special importance in the oil and gas industries. The NEL manages technology transfer programmes for the DTI in open systems, as well as operating the National Wind Turbine Centre.

—The National Physical Laboratory (NPL) is Britain's national standards laboratory, with responsibility for maintaining national measurements for physical quantities. It supplies essential calibration and technology transfer services for industry, and undertakes research on standards for engineering materials and information technology. The NPL is also the base for the

National Measurement Accreditation Service, which provides accreditation of calibration and testing laboratories in both the public and private sectors.

The three laboratories are currently executive agencies. However, the Government has reviewed their operations and decided that their future lies in the private sector:

—the LGC is to be established as a independent non-profit-making private sector company by the end of 1995–96;

—the NEL is to be sold in the second half of 1995; and

—proposals are to be invited for a management contractor to take over the NPL.

Technology Transfer

The DTI concentrates support on helping firms gain access to, and assimilate, technology; building up partnerships among firms; working with higher education; learning from overseas; and ensuring that firms have the facilities to enable them to innovate and to do business better. It is significantly improving companies' access to science and technology by making innovation services available locally—for example through Business Links (see p. 29). Many local innovation networks already exist which bring together expertise in higher education institutions, technical colleges, industrial research organisations and other bodies. The DTI intends to build on these networks and improve access to them for small and medium-sized businesses.

The DTI Innovation Unit works predominantly through the use of secondees from industry in six key areas covering both government and the private sector to stimulate wealth creation by encouraging a climate for innovation:

—encouraging the spread of best practice of innovation;

—commercial exploitation of the science base;

—improving industry–investor relations;

—encouraging innovation in education;

—improving the public understanding of innovation; and

—encouraging innovation in government.

DTI initiatives to enhance companies' awareness of management best practice will continue through the 'Managing in the 90s' programme. In addition, the Environmental Technology Best Practice Programme aims to improve industrial competitiveness and reduce pollution. The DTI is developing a more comprehensive information service about technology overseas. The DTI-supported Teaching Company Scheme (TCS) provides industrially relevant training for young graduates while they undertake key technology transfer projects in companies under joint supervision of academic and company staff; nearly 600 graduates are working on TCS programmes.

Ministry of Defence

Ministry of Defence provision for R & D in 1994–95 was £2,315 million. About £631 million of this was for medium and long-term applied research relevant to military needs. Much of this research is carried out in the MoD's Defence Evaluation and Research Agency (DERA). With the ending of the Cold War, the Government is committed to achieving a gradual reduction in real terms in spending on defence R & D.

In 1991 several research establishments merged to form the Defence Research Agency, which was expanded to form the DERA in April 1995. It is the largest single scientific employer in Britain. Its role is to provide scientific and technical services primarily to the Ministry but also to other government departments.

The DERA subcontracts about £200 million of research to industry and universities, ensuring that their know-how is harnessed to meeting military requirements. It also works closely with industry in order to ensure that scientific and technological advances are taken forward at an early stage into development and production. This technology transfer is not just confined to the defence industry but has also led to important 'spin offs' from defence into civil markets, in fields ranging from new materials and electronic devices to advanced aerodynamics. The latter in particular has been instrumental in giving Britain a leading role in civil aircraft design.

Recent innovations by the DERA include the development of an advanced composite rotor blade for the Lynx helicopter; and the invention of a special form of highly porous silicon, which can be made to emit light when irradiated with ultra-violet light and which opens up the technology of silicon opto-microelectronics.

Other Government Departments

Ministry of Agriculture, Fisheries and Food

MAFF co-ordinates its research programme with The Scottish Office Agriculture and Fisheries Department, the Department of Agriculture for Northern Ireland and the research councils, particularly the BBSRC (see p. 21). It also covers the research interests of the Welsh Office Agriculture Department.

Its research programme reflects the Ministry's wide-ranging responsibilities for protecting and enhancing the rural and marine environment; protecting the public, especially in food safety and quality, flooding and coastal defence, and animal health and welfare; and improving the economic performance of the agriculture, fishing and food industries.

The budget for research expenditure in 1994–95 was £137 million. Research is contracted with research councils, the Ministry's agencies, non-departmental public bodies, higher education institutions and other organisations.

Department of the Environment

The Department of the Environment funds research in several policy areas: environmental protection, including radioactive substances; water; the countryside; planning and inner cities; local government; housing; building and construction; and energy efficiency. The three largest sub-programmes are those on pollution-related climate change, regional and urban air quality, and the safe disposal of radioactive waste. Total research expenditure in 1994-95 was estimated at £96 million.

Department of Health

The Department of Health has developed an R & D programme for the National Health Service (NHS) which will provide a scientific basis for the promotion of health and the provision of health care. A key aim is to increase the effectiveness of the NHS by creating a sound base from which strategies in health care, operational policy and management can be defined and practice improved. Expenditure on R & D in the NHS was £437 million in 1994–95, representing 1.2 per cent of NHS expenditure. The Department also manages a Health and Personal Social Service research programme concerned with the needs of ministers and policy-makers, with emphasis on improving efficiency.

In December 1994 the Secretary of State for Health announced a number of measures to support research in the NHS:

—a new funding mechanism for NHS R & D;

—the intention to raise R & D funds by a levy on purchasers of health care;

—an extra £8 million which will be available in 1995–96 for research commissioned by the NHS;

—a new role for the Central Research and Development Committee in advising the NHS on how to invest its funds for R & D; and

—the creation of a National Forum to bring together the major sources of funds for health-related research (including the MRC, charities, universities and industry) to provide advice to the NHS and the Government.

Overseas Development Administration

The Overseas Development Administration (ODA) commissions and sponsors research on topics relevant to those geographical regions designated as the primary targets of the aid programme and of benefit to the poorest people in those countries. Expenditure on R & D in 1994–95 was estimated at £92 million.

The ODA's support for R & D is organised into five main programmes covering renewable natural resources; engineering-related sectors (water and sanitation, energy efficiency and geoscience, urbanisation and transport); health and population; economic and social development; and education.

R & D is also carried out as part of Britain's bilateral aid to particular countries. It draws upon a range of professional expertise of agencies such as the Transport Research Laboratory, the British Geological Survey and the Institute of Hydrology. The ODA contributes towards international centres and programmes undertaking R & D aimed at solving problems faced by developing countries. It also contributes to science and technology through the European Union. The EU's Science and Technology for Development programme seeks to stimulate simultaneous study in various parts of the world on specific scientific issues in health, nutrition and agriculture, which can contribute to progress in all developing countries.

The Scottish Office

The Scottish Office both contracts and undertakes itself a wide range of R & D commissions. Total R & D planned expenditure in 1994–95 was £73.3 million in support of its policy responsibilities, especially on agriculture, fisheries, health, the environment, education and home affairs. In some areas of work—medicine, agriculture and biological sciences, fisheries and marine science—research in Scotland has an international reputation.

Space Activities

Britain's support for civil space research is co-ordinated by the British National Space Centre (BNSC), a partnership between various government departments and research councils. BNSC encourages industry to exploit opportunities in space, based on appraisal of project costs and potential technological and commercial benefits. Through BNSC, Britain spent around

£194 million on space activities in 1994–95, mainly with DTI and research council financing. Around 66 per cent—£125 million—was devoted to programmes shared with the European Space Agency (ESA). The remainder supported R & D in industry, higher education institutions, government establishments and industry.

Higher Education Institutions

The higher education funding councils in England, Scotland and Wales are the largest single source of finance for higher education institutions in Great Britain. About 40 per cent of research carried out in higher education is financed from resources allocated by these bodies. These funds pay for the salaries of permanent academic staff who usually teach as well as carry out research, and contribute to the cost of support services. In Northern Ireland higher education institutions are funded by the Department of Education for Northern Ireland. The level of research performance of departments is a key element in the allocation of funding.

The research councils support research in higher education institutions in two main ways. First, they provide postgraduate awards to over 25 per cent of postgraduate students in science and technology. Secondly, they give grants and contracts to institutions for projects, particularly in new or developing areas of research, and for support units and facilities for research. The research councils have become responsible for meeting all the costs of the research projects they support in higher education institutions, except for costs associated with salaries of permanent academic staff and costs of premises and infrastructure.

The other main channels of support for scientific research in higher education are government departments, charities and industry. The European Union also provides substantial funding. Institutions are expected to recover the full cost of short-term commissioned research from the Government and industry.

The high quality of research in higher education institutions, and their marketing skills, have enabled them to attract more funding from a larger range of external sources, especially in contract income from industry. Co-operation continues between higher education institutions, industry and the Government in joint projects.

Institutions undertaking research with the support of research council grants have the rights over the commercial exploitation of their research, subject to the prior agreement of the sponsoring research council. They may make use of the expertise of the British Technology Group (see p. 42) to patent and license their inventions.

Science Parks

Science parks are partnerships between higher education institutions and industry to promote commercially focused research and advanced technology. In 1994 there were 46 such parks in operation, generally at or near universities. The parks accommodate over 1,200 companies, with an estimated combined annual turnover of over £1,000 million, and the companies employ around 20,000 staff. Most companies are engaged in work on computing, electronics, instrumentation, robotics, electrical engineering, chemicals and biotechnology. Research, development and training activities are the most common activities, rather than large-scale manufacturing. The biggest science park is in Cambridge, with about 85 companies on site. Recently opened parks are situated at Oxford, Cranfield, Westlakes (Cumbria) and York.

A growing number of universities offer industry interdisciplinary research centres. These include access to analytical equipment, library facilities and worldwide databases as well as academic expertise.

Other Organisations

Industrial Research and Technology Organisations

Research and Technology Organisations (RTOs) are independent organisations carrying out commercially relevant contract research and other services on behalf of industry, often relating to a specific industrial sector. Britain has the largest RTO sector in Europe, consisting of approximately 70 organisations, which together employ over 10,000 people.

These research organisations conduct confidential strategic research which companies need for their next generation of products or services. They may develop new products and carry out routine testing, particularly relevant to small companies which may not be able to afford the advanced instruments needed to make detailed measurements.

Charitable Foundations

Medical research charities provide a major source of funds for biomedical research in Britain. Their combined contribution in 1992–93 was about £320 million. The three largest contributors were the Wellcome Trust—the world's largest medical charity—with a contribution of £77 million, the Imperial Cancer Research Fund (£50 million) and the Cancer Research Campaign (£44 million).

British Technology Group

The British Technology Group (BTG) is among the world's leading technology transfer companies. It promotes the profitable commercialisation of technology by:

—developing and protecting technology arising from research carried out by individuals, higher education institutions and other research organisations which it considers will be commercially viable;

—licensing the resulting intellectual property rights to companies throughout the world; and

—assessing the commercial potential of companies' proprietary technology and licensing this technology to other companies worldwide.

The BTG administers about 10,000 patents covering around 1,400 technologies; in 1993–94 turnover was £29 million. The Group is owned by a consortium consisting of BTG management, employees and financial institutions.

Professional Institutions and Learned Societies

There are numerous technical institutions and professional associations in Britain, many of which promote their own disciplines or the education and professional well-being of their members.

The Council of Science and Technology Institutes has seven member institutes representing biology, chemical engineering, chemistry, food science and technology, geology, physics and hospital physics. The Engineering Council promotes the study of all types of engineering in schools and other organisations, in co-operation with its 210 industry affiliates, which include large private sector companies and government departments. Together with 41 professional engineering institutions, the Council accredits courses in higher education institutions. It also advises the Government on a range of academic, industrial and professional issues.

More than 300 learned societies play an important part in promoting science and technology through meetings, publications and sponsorship.

Royal Society

The most prestigious learned society is the Royal Society, founded in 1660. It has over 1,100 Fellows and more than 100 Foreign Members. Many of its Fellows serve on governmental advisory councils and committees concerned with research. The Society's estimated net expenditure on science and technology in 1994–95 was about £22 million. Almost 80 per cent of funding is derived from the Government, the remainder coming from private sources.

The Society encourages scientific research and its application through a programme of meetings and lectures, publications, and by awarding grants, fellowships and other funding. It recognises scientific and technological achievements through election to the Fellowship and the award of medals and endowed lectureships. As the national academy of sciences, it represents Britain in international non-governmental organisations and participates in a variety of international scientific programmes. It also facilitates collaborative projects and the exchange of scientists through bilateral agreements with academies and research councils throughout the world. It gives independent advice on scientific matters, notably to government, and represents and supports the scientific community as a whole.

The Society is increasingly active in promoting science understanding and awareness among the general public, as well as science education. It also supports research into the history of scientific endeavour.

Royal Academy of Engineering

The national academy of engineering in Britain is the Royal Academy of Engineering, founded in 1976. The Academy, which has about 970 Fellows, 57 Foreign Members and 12 Honorary Fellows, promotes the advancement of engineering for the benefit of the public. It receives a grant from the OST amounting to £2.6 million in 1995–96 and plans to raise over £5 million from other sources. Among the objectives of its programmes are the attraction of first-class students into engineering, raising awareness of the importance of engineering design among undergraduates, developing links between industry and higher education, and increasing industrial investment in engineering research in higher education institutions.

Other Societies

In Scotland the Royal Society of Edinburgh, established in 1783, promotes science by awarding scholarships, organising meetings and symposia, publishing journals and awarding prizes. It also administers fellowship schemes for post-doctoral research workers.

Three other major institutions publicise scientific developments through lectures and publications for specialists and schoolchildren. Of these, the British Association for the Advancement of Science (BAAS), founded in 1831, is mainly concerned with science, while the Royal Society of Arts, dating from 1754, deals with the arts and commerce as well as science. The

Royal Institution, founded in 1799, also performs these functions and runs its own research laboratories.

Zoological Gardens

The Zoological Society of London, an independent scientific body founded in 1826, runs London Zoo, which occupies about 15 hectares (36 acres) of Regent's Park, London. The Society is responsible for the Institute of Zoology, which carries out research in support of conservation. The Institute's work covers a wide range of topics including ecology, animal behaviour, reproductive biology and conservation genetics.

Whipsnade Wild Animal Park near Dunstable (Bedfordshire) is also managed by the Society. Other well-known zoos include those at Edinburgh, Bristol, Chester, Dudley and Marwell (near Winchester).

Many zoos play a vital role in programmes for reintroducing species of international importance (see p. 86).

Botanic Gardens

The Royal Botanic Gardens, founded in 1759, cover 121 hectares (300 acres) at Kew in west London and a 187-hectare (462-acre) estate at Wakehurst Place, Ardingly, in West Sussex. They contain the largest collections of living and dried plants in the world. Research is conducted into all aspects of plant life, including physiology, biochemistry, genetics, economic botany and the conservation of habitats and species. Kew has the world's largest seed bank for wild origin species and is active in programmes to return endangered plant species to the wild. It participates in joint research programmes in 52 countries.

The Royal Botanic Garden in Edinburgh, founded in 1670, is a centre for research into taxonomy (classification of species), for the conservation and study of living plants and for horticultural education.

Scientific Museums

The Natural History Museum, in South Kensington (London), which includes the Geological Museum, has about 67 million specimens in its reference collections. Specimens range in size from a blue whale skeleton to minute insects. It is one of the world's principal centres for research into natural history, offering an advisory service to institutions all over the world. The museum is keen to promote public appreciation of nature conservation and in 1991 opened a major exhibition on ecology, illustrating the diversity of living species. Some 1.75 million people visited the museum in 1993–94.

The Science Museum, also in South Kensington, promotes understanding of the history of science, technology, industry and medicine. Its extensive collection of scientific instruments and machinery is complemented by interactive computer games and audio-visual equipment for visitors to use. In this way the museum explains scientific principles to the general public and documents the history of science, from early discoveries to space-age technology. Educational facilities at the museum are being enhanced; around 300,000 schoolchildren a year go to the museum on organised school visits. Visitors to the museum totalled 1.3 million in 1993–94.

Other important collections include those of the Museum of Science and Industry in Birmingham, the Museum of Science and

Industry in Manchester, the Museum of the History of Science in Oxford, and the Royal Scottish Museum, Edinburgh.

Promoting Public Awareness

Many of the learned societies and other institutions, such as the Royal Society and the Royal Institution, work to improve the understanding and communication of science. They publicise scientific developments through lectures and publications for specialists and schoolchildren. Press conferences are held to provide the media with a steady flow of information about scientific developments.

The Committee on the Public Understanding of Science (COPUS), set up in 1986 by the Royal Society, the BAAS and the Royal Institution, acts as a focus for a wide-ranging programme in Britain to improve public awareness of science and technology. It recommends initiatives to its sponsoring bodies and promotes activities of other organisations and institutions. The annual COPUS Science Book Prize is awarded to authors of non-specialist books.

The Government has launched a national campaign (see p. 12) to increase the general level of understanding of science and technology issues throughout society.

Science Festivals

Science festivals are a growing feature of local co-operative efforts to improve public understanding of the contribution made by science to everyday life. Schools, museums, laboratories, higher education institutions and industry contribute to a broad range of special events.

The largest science festival in a single city in the world is the Edinburgh International Science Festival. In 1994 it included 16 exhibitions, 36 workshops and over 200 lectures; more than 200,000 people attended.

International Collaboration

Britain plays a key role in major international scientific facilities and research programmes.

European Union

Since 1984 the EU has run a series of R & D framework programmes in several strategic sectors, with the aim of strengthening the scientific and technological basis of European industry and improving its international competitiveness. The Third Framework Programme lasted from 1990 to 1994, with a budget of 6,600 million ECUs (£5,079 million).

Under the Fourth Framework Programme for Research and Technological Development, which runs from 1994 to 1998, about 12,300 million ECUs (£10,200 million) is available for expenditure on R & D. Britain played a significant part in shaping its structure and priorities.

Britain helped to secure agreement that the EU should develop 'generic' technologies with a large range of industrial applications, rather than just funding research projects to meet the needs of specific industrial sectors. The EU has also agreed that more resources should be devoted to disseminating technology from research projects to small and medium-sized enterprises.

Examples of the many EU research activities involving British organisations include the following programmes:

—Britain contributes to the EU Nuclear Fusion Programme, both by supporting research into fusion carried out at Culham,

Oxfordshire, and by hosting the Joint European Torus (JET) project, also based at Culham. Countries participating in JET include the EU members and Switzerland.

—The EU's Information Technology Programme, formerly known as ESPRIT (European Strategic Programme for Research in Information Technology), is part of the Fourth Framework Programme. It is a shared-cost collaborative programme, designed to help build the services and technologies that underpin the 'information society', with a strong emphasis on the needs and users of the market. The programme is open to companies, academic institutions and research bodies. Britain is currently participating in 369 projects.

Other International Activities

Over 550 British organisations have participated in EUREKA, an industry-led scheme to encourage European co-operation in developing and manufacturing advanced technology products with worldwide sales potential. Britain is currently engaged in 257 projects. Major projects include research into a European standard for high-definition television and into semi-conductors. There are 33 members of EUREKA including the 15 EU countries and the European Commission.

The COST (European Co-operation in the Field of Scientific and Technical Research) programme encourages co-operation between national research organisations across Europe, with participants coming from industry, academia and research laboratories. Transport, telecommunications and materials have traditionally been the largest areas supported. New areas include physics, neuroscience, and forests and forestry products. There are currently 25 member states. Britain takes part in 84 of the 102 current COST projects.

Britain is a member of CERN, the European Laboratory for Particle Physics, based in Geneva. Scientific programmes at CERN aim to test, verify and develop the 'standard model' of the origins and structure of the universe. There are 19 member states. The subscription to CERN is paid by the PPARC. In 1994 agreement was reached on the project to build a new Large Hadron Collider (LHC) particle accelerator. This is the next major step in particle physics research, and will enable a large variety of experiments in particle physics to be carried out. The LHC will be built in two phases; in the first phase a particle collider should be ready to begin experiments in 2004.

Contributions to the high-flux neutron source at the Institut Laue-Langevin and the European Synchrotron Radiation Facility, both in Grenoble, are paid by the EPSRC. The PPARC is a partner in the European Incoherent Scatter Radar Facility (EISCAT) within the Arctic Circle, which conducts research on the ionosphere.

Through the MRC, Britain participates in the European Molecular Biology Laboratory (EMBL), based in Heidelberg, Germany. Britain has been chosen as the location for the European Bioinformatics Institute, an outstation of the EMBL. The Institute, based in Cambridge, provides up-to-date information on molecular biology and genome sequencing for researchers throughout Europe.

The MRC pays Britain's contribution to the Human Frontier Science programme, on behalf of all the research councils. The programme supports international collaborative research into brain function and biological function through molecular level approaches.

The NERC's James Rennell Centre in Southampton is a focal point for Britain's contributions to the World Ocean Circulation Experiment, one of the major components of the World Climate Research Programme. The NERC also pays Britain's subscription to the Ocean Drilling Program (ODP). In 1993 Britain was chosen as the location for the scientific planning office of the ODP's Joint Oceanographic Institute for Deep Earth Sampling. The office is based at the University of Wales, Cardiff, for two years from October 1994.

Britain is a member of the science and technology committees of international organisations such as the Organisation for Economic Co-operation and Development (OECD) and the North Atlantic Treaty Organisation (NATO). It has taken part in the OECD's Megascience Forum set up to look at global planning of large facilities and programmes, and various specialised agencies of the United Nations.

Among non-governmental organisations, the research councils, the Royal Society and the British Academy were founder members of the European Science Foundation in 1974 and play a full part in its activities. The research councils also maintain, with the British Council, a joint office in Brussels to further European co-operation in research. Staff in British Embassies and High Commissions promote contacts in science and technology between Britain and overseas countries and help to inform a wide range of organisations in Britain about science and technology developments overseas. There are science and technology sections in Paris, Bonn, Washington, Tokyo and Moscow.

The British Council, which works in 228 towns and cities in 109 countries, promotes educational, cultural, scientific and technological collaboration between Britain and other countries. It works closely with the British scientific community, fostering British expertise by:

—supporting partnerships in science and medicine;

—promoting links and joint research collaboration and training;

—supporting overseas development, particularly human resource and institutional development;

—providing information on British research, education and training services; and

—demonstrating the excellence of British science, technology and engineering.

Major Research Projects

A substantial number of scientific advances have been made by British researchers. International collaboration is increasingly important, particularly in large-scale research projects, and many achievements have been made in partnership with overseas counterparts. A selection of recent major projects is given in this section.

Physical Sciences

Astronomy

Britain has a distinguished tradition in astronomy stretching back over 300 years to the pioneering work of the cataloguing of star positions for maritime navigation.

Astronomy and astrophysics concentrate on understanding how stars and galaxies formed and have evolved, and on working back to characteristics of the 'Big Bang', as well as testing the laws of physics under conditions totally inaccessible to laboratory experimentation. Britain remains at the forefront of this research, a major reason being the continuing development of excellent and innovative instruments.

In optical astronomy exciting discoveries have been made with the PPARC's network of telescopes at home and abroad, both ground-based and in space. The PPARC is a principal partner in the international La Palma Observatory in the Canary Islands. The observatory's four telescopes include the 4.2-metre William Herschel, the third largest single-mirror optical telescope in the world. On Mauna Kea, Hawaii, the PPARC has a 3.8-metre infra-red telescope, the largest telescope in the world designed specifically for infra-red observations, and the 15-metre James Clerk Maxwell radio telescope (built in collaboration with The Netherlands and Canada).

In 1992, using a new detector on the Mauna Kea infra-red telescope, a team of British and US astronomers discovered a rare form of hydrogen associated with auroras, H_3^+, around the polar regions of Saturn. Monitoring the strength of H_3^+ radiation may provide information on the interaction of the planet with wind particles from the Sun.

Through the PPARC Britain is taking part in the Gemini project (see p. 20). This is an international collaboration to build two 8-metre telescopes that will give all-sky coverage from sites in Hawaii and Chile.

Radio astronomy telescopes based in Britain complement the work of British optical, infra-red and millimetre wave telescopes on overseas sites as well as the British and ESA (European Space Agency) space programmes. Manchester University's Nuffield Radio Astronomy Laboratory at Jodrell Bank, where work on astronomy began in 1945, is one of the world's leading centres for radio astronomy. The earliest work at Jodrell Bank was on the study of meteors using radar techniques developed during the Second World War. This led to the discovery of new meteor streams and their identification with debris spread around the orbits of comets. The radar work confirmed that all meteors were members of our solar system.

Once the principles of radio astronomy were established, the director of Jodrell Bank, Sir Bernard Lovell, and his colleagues, gradually shifted their attention from meteors to cosmic radio waves coming from the Milky Way and beyond. This work grew to dominate activities at Jodrell Bank, and helped to establish radio astronomy as a revolutionary method of investigating the universe.

For projects which need high resolving power (the ability to see fine detail), Jodrell Bank scientists have adopted a method—invented by Sir Martin Ryle and colleagues at Cambridge University in the 1950s—of making small radio telescopes imitate much larger telescopes by linking them electronically. The MERLIN (Multi-Element Radio-Linked Interferometer Network) is Jodrell Bank's array of seven observing stations that together form a powerful telescope with an effective aperture of over 230 kilometres (140 miles). The MERLIN project was formally completed in 1990 and has quickly produced significant results. In 1991, for example, the MERLIN network observed patterns in the sky which point to the existence of dense clumps of matter at a very early stage in the evolution of the universe. This has advanced the study of gravitational lenses, predicted by Einstein's General Theory of Relativity.

Recent Astronomical Discoveries

The brightest known object in the universe, believed to be a violently active galaxy, was discovered in 1991 by astronomers from Cambridge University. A special machine was used to measure photographs taken from the UK Schmidt Telescope in Australia. This is the second most distant known object in the universe, but much brighter than the others—being 10,000 times more luminous than the Milky Way galaxy—and 300 million million times more luminous than the Sun.

A new X-ray star was discovered in 1989, with a space-borne instrument invented at Leicester University. It was later observed by the Royal Greenwich Observatory to be an ordinary low-mass star (less massive than the Sun) orbiting and losing mass to a compact companion object—possibly a black hole.

Black Holes

Black holes are collapsed stars where gravity is so strong that light cannot escape. To find black holes, astronomers look at binary systems in which a star less massive than the Sun is losing material to a companion neutron star or black hole.

In 1993, using new infra-red technology, astronomers at Keele and Oxford Universities 'weighed' the compact stellar object in an X-ray binary (double-star) system and found it to be a black hole of 16 solar masses. The British astronomers developed a new method of determining the mass of compact stars in these binary systems. These new observations make it possible to determine the mass of the compact star with a considerable amount of precision. The astronomers believe that the identification of black holes is no longer in doubt and that it will now be possible to study them and their influence on their surroundings in some detail.

Space Activities

Britain's support for civil space research is co-ordinated by the British National Space Centre (BNSC) (see p. 37), a partnership between various government departments and research councils.

The major part of Britain's space programme is concerned with satellite-based Earth observation (remote sensing) for commercial and environmental applications. Remote sensing involves using instruments mounted on satellites and aircraft to study the Earth's land, sea, ice and atmosphere from space. It has considerable potential commercial applications, including mapping, crop monitoring, oil and mineral exploration, fisheries, weather forecasting and monitoring iceberg movements.

Britain has committed around £70 million to the ESA's ERS-1 satellite, which was launched in 1991. ERS-2, to which Britain has contributed £45 million, was launched in April 1995. British firms provided two of the instruments on the satellites: a synthetic aperture radar (SAR) and an along-track scanning radiometer (ATSR). The SAR is capable of providing high resolution images of the Earth with 24-hour coverage, irrespective of cloud cover conditions. ATSR measures global sea surface temperature to a very high degree of accuracy. British remote-sensing instruments are also being flown on other satellites including the Upper Atmosphere Research Satellite, operated by the United States National Aeronautics and Space Administration (NASA).

The Earth Observation Data Centre at Farnborough (Hampshire), operated by the National Remote Sensing Centre Ltd, processes the data output from ERS-1 and other remote-sensing spacecraft, and will do so for successor missions. It is also one of the ESA's four processing and archiving facilities which store and distribute remotely sensed data for both scientific and commercial purposes.

Britain is contributing £284 million to the Polar Platform and ENVISAT-1 programmes. The ENVISAT payload, carrying a new generation of ESA environmental instruments, will fly on the British-led Polar Platform. For ENVISAT, Britain is leading the development of an advanced action microwave instrument, and an advanced version of the precision instrument to measure infra-red emissions over land and sea.

In space science, Britain participates in all of the ESA's missions. These range from participation in the NASA Hubble

Space Telescope, for which British Aerospace built the solar panels, to the Ulysses solar polar probe, which is the first spacecraft to overfly the poles of the sun. The Hubble Space Telescope, launched in 1990, carries British-built solar arrays and the Faint Object Camera, both provided by the ESA. Despite the 'spherical aberration' in its optics which initially affected its focusing power, the telescope is now producing images of great clarity, and British astronomers are active in its use. The ability to resolve very faint objects at great distances is contributing significantly to understanding the nature and origins of the universe.

Britain is contributing substantially to the Cluster and SOHO missions to study the Sun, the Earth's magnetosphere and the solar wind, and to the Infrared Space Observatory, all due for launch in 1995. It is also participating in the ESA's X-ray spectroscopy mission, due for launch in 1999.

There are bilateral agreements for space research between Britain and other countries, such as the United States through NASA, and Japan. British groups have been involved in developing, for example, the widefield camera for Germany's X-ray satellite ROSAT, a spectrometer for the Japanese-built Yohkoh satellite and an X-ray telescope for the Russian Spectrum-X mission.

In Europe, Britain is both a leading producer and user of satellite communications technology, exploiting commercially expertise developed within the ESA's satellite communications programmes.

Nuclear and Particle Physics

Research in particle physics and nuclear structure physics (the study of the elementary constituents of matter and their constituents) is supported by both the EPSRC and the PPARC and co-ordinated by the CCL.

In recent years British physicists have made impressive advances in nuclear structure research, including the original discovery of superdeformed, rapidly spinning nuclei; the discovery of new examples of proton radioactivity; and important studies of the clustering of nuclei—the so-called nuclear module. Recent achievements include theoretical progress to show how defects such as cosmic strings may leave distinctive signals in the cosmological microwave background, and the discovery that aluminium uptake in the body may be 100 times greater than previously thought.

The ISIS machine, the world's most intense source of pulsed neutrons and muons, is used by scientific teams from over 20 countries to probe a range of materials, such as high-temperature superconductors. In 1992 ISIS was used to perform around 600 experiments. The low-energy neutrons produced by ISIS are able to reveal geometrical details of atoms and molecules in widely used materials such as industrial polymers, liquid crystals, hydrogen storage alloys and anaesthetics. A test-bed for radioactive nuclear beams is being funded on the ISIS machine. This area of research is a priority topic in nuclear structure physics.

Eurogam

Experimental observation of spinning nuclei requires highly sophisticated equipment, consisting of detectors which measure the amount of gamma-ray energy emitted. British and French scientists have constructed Eurogam, the world's most advanced gamma ray spectrometer.

The scientists will initially use Eurogam's detector array to study nuclear matter under the most extreme conditions, such as nuclei rotating at extreme velocities. Eurogam will operate in conjunction with the Daresbury Recoil Mass Separator, a highly sensitive device for identifying species of nuclei which emit gamma rays. Together they will provide an unparalleled facility for studying the nucleus.

Synchrotron Radiation

The Synchrotron Radiation Source at Daresbury provides very high-energy radiation, from X-rays to the infra-red part of the electromagnetic spectrum. It is used by some 2,000 British and overseas scientists to investigate, among other things, the expression and suppression of genes and what happens when chemical 'messengers', large protein molecules, arrive at receptor sites in living cells; and what happens at the boundaries between two different materials in semiconductors.

The world's first continuously variable wavelength, ultraviolet microscope has been installed at Daresbury. Funded by the EU, it will be used to obtain 'optical slice' images through biological and other materials.

Particle Physics

Particle physics experiments require the use of large colliders and accelerators. They form the basis of a number of international collaborations, performed at overseas laboratories, particularly CERN, the European Laboratory for Particle Physics in Geneva.

The current CERN programme is focused on the Large Electron-Positron Collider, the world's biggest operating particle collider, with British physicists and engineers collaborating in the design and construction of three of the four highly sophisticated experiments currently in progress. British physicists are playing an important role in designing experiments for the new Large Hadron

Collider (LHC) to be built at CERN (see p. 51). They are also collaborating in the two major experiments at HERA, the electron-proton collider at the Deutsches Electronen Synchrotron (DESY) Laboratory in Germany; both are designed to explore the internal structure of the proton and its constituents.

Scientists at St Andrews University are to build a 'quasi-optical' electron spin resonance spectrometer, a radical new type of spectrometer able to observe chemical, biological and physical reactions that are too swift and faint to be seen with existing instruments. Electron spin resonance (ESR) is a technique with important applications in medical research and the development of new drugs. It is also useful in fields ranging from forensic science to studies of new materials like high temperature superconductors.

Electronics and Communications

British scientists, research organisations and companies have been responsible for some of the most important inventions in radio, television and radar (see Appendix), which formed the basis for the development of the modern electronics industry. Products pioneered in British companies and elsewhere are often licensed to overseas manufacturers.

Current research areas cover opto-electronics, molecular electronics, advanced control technology, advanced silicon technology and design automation. Priority areas include software and supercomputing in engineering.

Electronic systems in computer-aided design, process control instrumentation and robotics are commonplace in manufacturing industry. Robotics are used extensively on production lines in the motor industry, and advanced electronics incorporated into the vehicles that are manufactured. Companies such as GEC and Oxford Instruments develop and produce analytical instruments for industrial research and advanced electronic medical equipment, including ultrasound scanners, patient monitoring systems and magnetic resonance imaging (see p. 97).

Radar was invented in Britain and British firms are still in the forefront of technological advances. Racal Avionics' X-band radar for aircraft ground movement control is in use at airports in several countries. Solid-state secondary surveillance radar, manufactured by Cossor Electronics, is being supplied to 50 overseas civil aviation operators. The world's first portable radar display unit, the Pilotwatch system, developed in Britain by the firm dB Electronics, will enable coastal policing and marine pilot navigation to be made more effective. The Pilotwatch system can broadcast radar pictures from any number of radar stations to any number of portable receivers.

Superconductors

New ceramic materials developed by GEC-Marconi may prove to be superior to existing high-temperature superconductors (materials with low electrical resistance). Patent applications have been filed for three materials based on cadmium, lead and copper oxides, which have been found to be both stable and reproducible. The research was part of a multinational EU project dedicated to developing superconducting cables and conductors operating at liquid nitrogen temperatures and above.

By the end of the 1990s further developments are expected to lead to applications in power generation and transmission, electromagnetic stores for off-peak electricity and transport systems using magnetic levitation. The world's highest-temperature superconductor is currently being produced by a research unit at Cambridge University.

Computing Services

Britain has made important contributions in opto-electronics (the combination of optics and electronics) and parallel computing, which is based on transputers working in parallel to give the same capacity as a supercomputer. In 1983 Inmos produced the transputer, which combines all the main functions of a simple computer on a single chip.

British companies are in the forefront of software development; Kerridge Network Systems, for example, devises programs for portable computers. British firms and research organisations, with government support, have been engaged in the development and application of the new family of 'three-five' semiconductor materials (such as gallium arsenide). In 1984 GEC released the first commercially available gallium arsenide microprocessor, which can operate at much higher frequencies than the common silicon type. The Trax, a novel design of supercomputer processor with a wide range of applications, including high-definition television and defence systems, was originally developed at Brunel University, near London.

Major advances are being made by British firms in the field of 'virtual reality', a three-dimensional computer simulation technique with a wide variety of industrial and other applications. The Advanced Robotics Research Centre in Salford is conducting research into virtual reality technology. Virtual reality is being used to design buildings and a range of products, including cars, pharmaceuticals and machine tools.

Networking

BT and the academic community are collaborating on the development of SuperJANET (the Super Joint Academic Network), a new high-speed fibre-optic network which will link computer systems in higher education and research institutions in Britain. SuperJANET will complement the Joint Academic Network (JANET) which currently links 200 sites and has served universities and research institutes with large computer power for more than a decade. The SuperJANET project aims to transform JANET into a 'multiservice' network.

Financed by the higher education funding councils, SuperJANET can transmit information much more quickly than previously, transmitting the equivalent of a 5,500 page report in less than one second. The increased power of SuperJANET comes from the use of fibre-optic links between computers and users, allowing very large amounts of data, graphics and visual images to be flashed across the network. The new network will open up entirely new ways of storing, processing and distributing information. SuperJANET is expected to support a very wide range of teaching and learning activities.

Telecommunications

BT, Britain's largest telecommunications operator, has led in the development of optical fibre communications systems and has paved the way for simpler and cheaper optical cables by laying the first non-repeatered cable over 100 km long, and by developing the first all-optical repeater. BT's Integrated Services Digital Network (ISDN) enables users to transmit speech, data, graphics and video communications over a single line. In 1994–95 the company spent £271 million on R & D. Its research centre is based at Martlesham Heath, Suffolk.

BT is working on advanced optical and radio network technology for lower-cost communications, and on software engineering to provide better managed and more 'intelligent' national and global networks. It is employing speech, data and visual technologies in the production of the videophone, video-conferencing and teleworking. Up to 800 R & D projects are carried out by the company at any one time; another recent example is the development of a video handset, with an optical telephone link. Around two-thirds of BT's research is in software engineering.

Mercury Communications, BT's main competitor in Britain, also has an advanced fibre-optic network. Mercury One-2-One launched the world's first Personal Communications Network system in 1993.

GPT, Britain's largest telecommunications manufacturer, invests substantially in research. Much progress is being made in GPT's systems, voice processing, computer-assisted telephony, videotelephony and videoconferencing.

Britain has also pioneered the development of a standard for digital cordless telephony (CT2/CA1), which has been adopted by a number of countries in Europe and is attracting much interest from around the world.

Aerospace

Britain has led the world in many aspects of aerospace R & D over the last 80 years. Pioneering achievements involving British technologists and aviation specialists include radar, jet engines, Concorde, automatic landing, vertical take-off and landing, flight simulators, swing-wing military aircraft and ejector seats. About 13 per cent of Britain's investment in industrial R & D goes to the aerospace industries.

Most developmental work is carried out by industry, with some assistance from government research establishments. A substantial amount of long-term aerospace research and formal trials work is conducted in partnership with universities and industrial research centres. Most of the Government's spending on aerospace R & D is concerned with military programmes.

Among important recent technological advances in aviation equipment is 'fly-by-wire', devised by British Aerospace (BAe), with Marconi and Dowty Boulton Paul, in which flying control surfaces are moved by electronic rather than mechanical means. GEC-Marconi developed the world's first fibre-optically signalled ('fly-by-light') aircraft control system.

The concept of head-up display (HUD) was pioneered and developed in Britain. This system electronically projects symbols into the pilot's view, avoiding the need to look down at instruments. GEC-Marconi has also developed a holographic HUD, which enables pilots to fly at high speeds at very low altitude in darkness.

Aerospace companies such as BAe, Smiths Industries and GEC-Marconi carry out R & D on a vast range of equipment for civil and military aircraft, such as navigational equipment, sophisticated radar and air traffic control systems, and flight control and landing systems. BAe has introduced a Terrain Profiler Matching System, which compares aircraft height with a digital map store in its memory. GEC-Marconi's airborne computer can produce colour maps on cockpit electronic displays.

Companies such as Cossor, GEC, Plessey and Racal are world leaders in radar technology. Siemens Plessey and Racal are working on Microwave Landing Systems, set to replace the Instrument Landing Systems in current use.

Aerospace Engines

Rolls-Royce is one of the Western world's three leading manufacturers of aerospace engines. Its most important civil engine is the

RB211 turbofan, several versions of which are in operation, the RB211-524G and 544H being the latest to enter service. They feature Rolls-Royce's unique wide-chord fan blades, together with an electronic fuel control system, giving improved thrust and fuel efficiency. The RB211-544H's modular design allows it to be upgraded to incorporate new technology.

The production of larger-capacity aircraft means that there is a need for engines of 50,000 lb or more thrust. The latest Rolls-Royce engine—the Trent 800, the world's most powerful turbofan—is due to enter commercial service on the Boeing 777 in early 1996.

In 1993 Rolls-Royce announced two new research programmes for military aircraft. Funded jointly with the MoD, the first project will investigate and demonstrate ways of applying advanced technologies to combustors and high-pressure turbines for use in combat aircraft engines. The second will demonstrate advanced compression systems, low pressure turbines, reheat systems and nozzles. Both programmes, aimed at boosting the performance of combat planes in the future, also have 'spin off' for civil engine developments.

Life and Medical Sciences

Many far-reaching scientific advances are being made in Britain in medicine, genetics, biotechnology and pharmaceuticals. For example, Britain continues to make outstanding contributions to organ transplant surgery and medical diagnostic techniques. Recent discoveries include: the identification of genes linked to certain diseases such as cystic fibrosis, and the subsequent development of gene therapies; the development of artificial blood; and the development of a treatment for rheumatoid arthritis using humanised antibodies.

Britain has a particularly strong reputation in protein engineering and antibody research and many British research teams working in the life sciences have established international reputations.

Medicine

Some recent important medical developments funded by the Medical Research Council and examples of the work of cancer research charities are outlined below.

Genetics

As in other major medical research centres worldwide, the effort devoted to molecular genetics funded by the MRC has increased substantially in recent years, reflecting the growing importance of this area in the development of drugs, vaccines and diagnostic techniques. Britain is in the forefront of genetics research, and is one of the main countries engaged in the process of isolating and mapping the complete sequence of genes—the genome—in the cells of humans. The research, and the study of gene products, has vast potential for the understanding of human biology and many diseases. A major national research project at the MRC is the Human Genome Mapping programme, involving mapping of the complete sequence of 100,000 genes in a human cell.

Genome research in Britain is conducted in higher education institutions, MRC establishments and centres supported by other agencies. The MRC has also established a Human Genome Mapping Project Resource Centre, which provides services and reagents to British and European research communities, including to industry. The resource centre is located on a new 'genome campus' at Hinxton in Cambridge, which also houses the Sanger Centre—a large-scale sequencing factory jointly funded by the

Wellcome Trust and the MRC—and the European Bioinformatics Institute, an outstation of the EMBL (see p.51).

A further initiative in the genetic approach to human health will build on the Human Genome Mapping Project, encompassing all aspects of genetic diseases and their future treatment. Centres of expertise in gene therapy have been established in Birmingham, Edinburgh and London. The centres aim to treat patients with inherited disease by inserting a healthy copy of a faulty or missing gene into cells in the patient's body.

Gene Therapy for Cystic Fibrosis

In August 1993 the first trials began at the Royal Brompton National Heart and Lung Hospital in London, in which sufferers of the inherited disease, cystic fibrosis, were treated by gene therapy. The therapy, administered by spray into the nose and lungs, contains a normal functional copy of the defective gene, packaged in minute fat globules (liposomes) which fuse with the recipient's cell membrane. The gene then works inside the cell to produce the protein whose absence allows the lungs to fill with mucus and make the patient prone to infection.

The cystic fibrosis gene was identified and isolated in 1989. Scientists at Oxford and Cambridge Universities used genetic engineering techniques to produce mice with many of the characteristics of cystic fibrosis in humans. Using these mice as models, they injected a normal copy of the defective gene into their airways and restored them to normality. Scientists at the MRC Human Genetics Unit at Edinburgh developed another mouse model which mimicked more closely the disease characteristics. They then refined the gene complex into a spray that could be inhaled by sufferers of the disease.

Cancer

The prospect of injecting 'killer' genes into cancer cells is one of the most promising areas of medical research. Imperial Cancer Research Fund scientists have devised a gene therapy for malignant melanoma, a fast-increasing form of skin cancer. Human trials began at the Fund's Oxford Clinical Oncology Unit in 1993.

The first vaccine to protect specifically against cancer has been developed by British scientists funded by the charity Cancer Research Campaign (CRC). The result of more than 25 years research, and a total investment by CRC of £5 million, the vaccine will be used to protect against the Epstein Barr virus, which is associated with a number of different cancers. The MRC is supporting gene therapy for cancer at two of the three gene therapy centres which it has established.

The MRC supports basic molecular and cellular studies on cancer, its genetic predisposition, detection and diagnosis, and measures to improve treatments, including clinical trials. Complementing the work of the medical charities and the Department of Health, the MRC's strategies in the cancer field are focusing on the understanding of tumour biology.

Artificial Blood

An important medical advance has resulted from collaboration between scientists at the MRC's Laboratory of Molecular Biology and Somatogen Inc of the United States who, in 1992, produced an artificial form of haemoglobin, the oxygen-carrying red protein in human blood, that could serve as a blood substitute. Using genetic and protein engineering, unlimited amounts of haemoglobin can be produced free from human infectious agents. The team devised a substance that lasts at least three times as long in the bloodstream as earlier artificial materials, transports oxygen efficiently and, unlike earlier preparations, does not cause kidney damage.

The new blood substitute is undergoing clinical trials in the United States and is expected to become commercially available within five years.

Infections and Immunity

The field of immunity has seen radical advances following the application of techniques used in molecular and cell biology and protein engineering. Recent MRC research on the role of peptides during the processing of antigens is contributing to vaccine development. Current work is aimed at understanding competing strategies in the immune system and the apparent effectiveness of infectious agents in exploiting these mechanisms. Other work is concerned with identifying the vulnerable stages of parasite life cycles in order to develop new drugs and vaccines to treat schistosomiasis and malaria.

The MRC plays a vital role in the international research effort to combat human immunodeficiency virus (HIV) and AIDS. Work is directed at developing antiviral drugs and vaccines, and assessing the extent of the epidemic and behavioural factors of transmission.

A major long-term investigation into premature births started in Britain in 1993. Funded by the MRC, its purpose is to test the idea that premature labour is frequently triggered by mild infections in the mother that could be cleared up by antibiotics. Over 10,000 women in 100 centres in Britain and abroad are expected to be involved over four years—the largest trial of its kind worldwide. The study is being co-ordinated by doctors at hospitals in Dundee and Leicester.

Biotechnology

Biotechnology includes the manufacture of products using genetic modification techniques. The discovery of the structure of DNA at Cambridge University in the 1950s (see p. 1) formed the basis for these techniques.

Biotechnology is expected to have a major impact in many industrial sectors by the next century. Its most advanced commercial applications are in pharmaceuticals and healthcare, along with food, agriculture and the development of advanced materials such as engineering plastics. Britain has made major advances in developing drugs such as human insulin and interferons, genetically engineered vaccines, and in producing antibiotics by fermentation; alternative bactericidal drugs based on Nisin, a food preservative made in Britain; medical diagnostic materials like biosensors; and agricultural products such as crop varieties that are resistant to infection or have improved qualities.

Major companies, like ICI, Glaxo and SmithKline Beecham, undertake extensive research in biotechnology. ICI, for example, developed Biopol, the world's first bio-degradable plastic. Britain's 50 small independent biotechnology firms specialise in medical diagnostics, microbial pesticides, plant breeding, waste treatments and the production of specialised enzymes.

In 1985 Celltech became the first company in the world to produce bioactive human calcitonin. This is the hormone used in the treatment of Paget's disease, a weakness in bone structure. The disease can now be treated using bioactive materials that interact with human tissue to form healthy bone.

Celltech has pioneered the mass production of monoclonal antibodies—artificially cloned proteins which can seek out a particular substance in the body. The antibodies are used to diagnose diseases, identify different blood types and in the treatment of a range of conditions, including cancer.

A second generation of vaccines based on recombinant DNA technology includes SmithKline Beecham's Energix-B vaccine against hepatitis. Drugs are being developed that act on defective genes either in the human host or in the infecting organism.

Britain leads in the development of molecular graphics, which contribute to the rational design of new and improved medicines through a computer-aided technique for analysing the structures of complex organic molecules on a visual display unit.

In 1992 MRC scientists developed a treatment using humanised monoclonal antibodies which relieves rheumatoid arthritis symptoms for up to eight months. Using genetic engineering, they reshaped rat antibodies to appear like human antibodies, thus bypassing the body's rejection reaction. Further studies are in progress to lengthen the period of remission.

In 1991 Professor Alec Jeffreys of Leicester University—the inventor of DNA fingerprinting—announced that he had invented a digitalised version of the DNA fingerprinting test. The new genetic profiling technique will make DNA fingerprints more accurate and easier to distinguish. For the first time it provides a method whereby different forensic laboratories can exchange and compare information, enabling international DNA fingerprint databases to be set up. The new test, which is being marketed by Cellmark Diagnostics, will also reduce the time taken to produce the fingerprint from four weeks to two days.

Agriculture and Food Research

The government agriculture departments, the BBSRC and private industry share responsibility for research. This is carried out in a network of research institutes, specialist laboratories and experimental husbandry farms; with Horticulture Research International; and in higher education institutions.

Animal Science

BBSRC work on basic animal physiology and genetics at the molecular and cellular levels underpins more strategic research related to animal production, health and welfare. The Council supports genome mapping which will allow the genetic identification of a number of performance traits and heritable defects in farm animals. Council scientists are also responsible for co-ordinating an extensive European consortium which aims to produce a complete genetic map of the pig.

Recent developments have seen the transfer of transgenic technology to industry; in 1992 a pharmaceuticals company, Pharmaceutical Proteins, refined the production of the protein alpha-l antitrypsin in the milk of genetically altered sheep. This protein will be used in the treatment of emphysema.

Together with a British company, Animal Biotechnology Cambridge (Mastercalf), and the United States Department of Agriculture, BBSRC scientists have developed a technique for separating the X–(female) and Y–(male) carrying sperm of bulls. This has been used successfully to produce calves of predetermined sex.

Animal Disease Research

Using modern molecular biological techniques, research council and MAFF scientists have devised new vaccines against animal diseases of considerable international economic importance. These include:

—the Aro A live salmonella vaccine against salmonella in chickens;

—the capri-pox virus vaccine containing rinderpest viral genome, which protects against both lumpy skin disease and rinderpest in cattle; and

—a vaccine against mastitis in cattle.

A new diagnostic test for salmonella has also been devised by MAFF scientists and has received great interest worldwide, particularly by human diagnostic laboratories.

Food Science

Scientists at the BBSRC Institute of Food Research have discovered the structure of the pectin-degrading enzyme, pectate lyase, which will open up ways of modifying the use of pectin as a gelling agent and improving the process of fruit juice clarification. Its scientists have also completed a comprehensive genealogical analysis of the genus Bacillus. This research has relevance to the agro-food and pharmaceutical industries as well as to health.

Plant Science

BBSRC research studies the genetic improvement of crops, processes affecting crop yields and soil fertility, crop nutrition and protection, managed grassland ecosystems, lower input farming systems, possible alternative crops and novel, non-food uses for crops.

Significant advances have been made in the analysis of crop genomes, in the study of plant reproductive processes and plant/herbivore interactions, and in identifying and manipulating genes for plant products which can be used to alter pest behaviour.

Hybrids with enhanced drought tolerance have been produced and new disease resistance introduced into white clover using wide crossing and embryo rescue techniques.

The order of genes within chromosome segments of wheat and rice has been shown to be the same. This discovery means that the whole plant and molecular genetics of rice will help future strategies in wheat breeding and the isolation of important genes.

The Council's scientists have successfully isolated a gene required for flower production. They found that when the gene was activated by a mutation, plants were unable to produce flowers but continually produced shoots instead. This discovery may eventually enable scientists to modify the pattern of flowering in a controlled way.

Engineering

Recent advances in BBSRC research include the development of mathematical models of spray dispersion, which can be used to estimate spray drift and deposition patterns for a range of crop, spray and environmental conditions. Collaborative research is showing which patterns are biologically the most effective. In 1992 scientists developed a computer-aided patch spraying system: instead of spraying the whole field, this system targets only those areas where weeds are growing.

BBSRC-funded scientists have discovered that a minute worm (*Phasmarhabditis* sp) belonging to the roundworm family will attack and kill slugs which cause considerable damage to crops such as winter wheat and lettuce. Using the worm has considerable advantages over chemical pesticides since they attack only slugs, leaving other small creatures such as earthworms untouched. Fed on liquid food, the worms can be grown in vast quantities and then kept in a clay mixture under refrigeration until needed. They are now being produced commercially in a pilot plant by the Microbio division of the Agricultural Genetics Company.

Chemicals

Britain's chemicals industry is at the forefront of modern technology, spending the equivalent of 6 per cent of total sales on R & D. Pharmaceuticals is the most research-intensive sector of the industry (see p. 6), while a large proportion of the world's R & D in agrochemicals is conducted in Britain.

Research undertaken by the chemicals industry over the last few years has led to significant technological and commercial breakthroughs. ICI is at the forefront of global efforts to develop substitutes for chlorofluorocarbons (CFCs); in 1991 it opened the world's first plant to manufacture HFC 134a, a less damaging alternative to CFC, in Runcorn, Cheshire.

In agrochemicals, notable British discoveries include synthetic pyrethroid insecticides, ICI's diquat and paraquat herbicides, systemic fungicides and aphicides, and methods of encouraging natural parasites to attack common pests in horticulture.

In fibres, Courtaulds has invented a new solvent-spun, biodegradable fibre, Tencel, which is twice as strong as cotton but soft enough to be used by designers of luxury garments.

Pharmaceuticals

Britain is the world's fourth largest exporter of medicines. In 1992 the pharmaceuticals industry invested more than £1,450 million in R & D, representing about 8 per cent of total world spending on medicines research.

Research conducted by Zeneca, Glaxo, SmithKline Beecham and Fisons has led to the development of semi-synthetic penicillins and cephalosporins, both powerful antibiotics; new vaccines; and new treatments for asthma, migraine, coronary heart disease and arthritis.

Glaxo is Britain's biggest pharmaceuticals company. It has a new £700 million research centre at Stevenage (Hertfordshire), which opened in April 1995. The company manufactures Zantac, the world's best-selling medicine, which is used in the treatment of gastric ulcers. It has developed several other new drugs, including Zofran, an anti-nausea drug for countering the unpleasant side-effects of cancer treatments, and drugs to treat migraine that stimulate receptors in the brain. It is at present conducting research into drugs for controlling certain viral diseases and cancers by regulating malfunctioning genes.

SmithKline Beecham, which manufactures four of the world's top-selling antibiotics, has developed Augmentin for treating infections that have become resistant to antibiotics.

Wellcome's drug zidovudine (AZT) is still the only anti-viral agent for the treatment of HIV infection that is approved in Britain and overseas. Zoladex, a drug made by Zeneca to treat prostate cancer, is being developed to treat breast cancer and benign gynaecological disorders.

Britain now leads the world in artificial bone research. Government-funded researchers at London University have developed a material to mimic the mechanical properties of real bone. This new man-made 'artificial bone' is a composite of hydroxyapatite and polyethylene. It is estimated that an artificial hip made from this material will last for about 25 years. At present a patient can expect a metal artificial hip to last for around 15 years.

Environmental and Earth Sciences

Growing national and international pressures for higher environmental standards and the need to halt the degradation of the environment have resulted in greater emphasis on environmental research. Britain spends large sums on such research in both the private and public sectors.

While environmental research plays a part in the work of all the research councils, the lead agency is the Natural Environment Research Council, much of whose work is done in collaboration with major international programmes on global environment issues. British scientists play a major role in research on assessing and repairing environmental harm caused by pollution; there is a substantial R & D effort, supported by the Government, to improve the technology of pollution control. The Government has also launched a new Environmental Technology Best Practice Programme, which provides information to business about the cost-effective reduction of pollution and waste.

Earth observation (see p. 58) makes a vital contribution to many environmental programmes, both national and international, concerned with geology, land use, pollution control, weather and climate studies.

Global Environment

Global Atmospheric Modelling Programme

Using models generated by computers, climatologists can predict how a gas introduced into the atmosphere at one point is distributed in varying amounts around the globe. The NERC supports a consortium of five universities on a project to model the atmosphere using computers. This programme concentrates on the natural variability of the atmosphere and its response to 'greenhouse forcing', and the dynamics and prediction of ozone depletion. Research has begun into the processes controlling the life cycles of atmospheric pollutants in the stratosphere and pollution problems nearer the ground. Britain is thus contributing considerably to the vital task of trying to understand the factors involved in global warming and in the potentially damaging changes in the atmosphere.

Terrestrial Initiative in Global Environment Research (TIGER)
This NERC initiative, started in 1991, is the largest study of the natural environment yet undertaken in Britain and will make a significant contribution to a number of global change projects. TIGER focuses on the response of nature to effects such as global warming and stratospheric ozone-depletion at sites in Britain and in selected tropical sites. In 1993 it published new findings suggesting that plants in a future 'greenhouse' atmosphere may breathe out less water vapour than previously predicted. This could result in a drier world, with reduced cloud formation and less rainfall. TIGER's two research sites in Britain are complemented by others in Amazonia, Africa and Canada. Data from experiments on energy/water balance carried out in the Amazonian rainforest are now being used in global climate models.

Resources
The NERC carries out strategic surveys of Britain's land mass and offshore regions, which give essential information on mineral resources, water resources, environmental protection, natural hazards and land use. A comprehensive geological survey of Britain's inner continental shelf, completed by the British Geological Survey, has yielded a detailed picture of its margin and associated natural resources. Britain is the first country in the world to have completely mapped its offshore area in this way. A 15-year programme of onshore re-survey was begun in 1991 and has already revealed a new phase of mineral deposits in Cornwall.

The NERC's Offshore Survey and Geophysical Monitoring project investigates the mineral resources and seismic hazards of

Britain's offshore areas. Two deep-ocean seismic experiments were undertaken during 1992; the results will aid geological interpretation of former plate collision zones.

Ocean Drilling Program

Britain participates in the Ocean Drilling Program (ODP), the world's largest and most successful international scientific research project in the Earth sciences. Current studies involving British scientists include those on the geology of ocean basins, mid-ocean ridges, marine sedimentation and ocean circulation.

Recent work by NERC-supported scientists from drill sites off Western Canada has led to the discovery of a large metal sulphide deposit which has apparently formed on the sea floor. The results have changed accepted views of the processes determining the formation of ancient mineralised sulphide deposits and might be applied to exploration of land-based deposits, which would be of economic value.

Marine Sciences

The Biogeochemical Ocean Flux Study (BOFS), begun in 1988, has now been completed. The project looked at how carbon dioxide (CO_2) is absorbed by the oceans and buried in the sea floor sediments, helping to stabilise the greenhouse effect. Comparisons of recent results show a large variation in the amount of animal and plant debris reaching the sea floor. Half of this variation can be linked to changes in ocean circulation, demonstrating that biological variability in the open oceans is governed by environmental changes that can be studied in greater detail. Funds are now being made available for a new project on how plankton in the North Atlantic react to their environment and their role in climate change.

The NERC's Land Ocean Interaction Study (LOIS) is a major project to study the dynamics of coastal environments and the flow of various materials into and across the coastal zone from the air, the land and rivers, and the sea. The study will provide the basis for predictions of the impact of future natural and man-made changes on the coasts.

The Fine Resolution Antarctic Model (FRAM), set up by the NERC to simulate 16 years of ocean dynamics, has now been completed. Its discoveries are causing oceanographers to revise their thinking on how oceans transport heat around the Earth, an important step in climate prediction.

A new Ocean Circulation and Climate Advanced Modelling Project (OCCAM) is designed to improve global ocean models for research into climate change on timescales ranging from a few months to 50 years.

Polar Sciences

The British Antarctic Survey (BAS), one of the NERC's research institutes, has been undertaking research in Antarctica for many decades. It has five permanent research stations in Antarctica, carrying out continuous research and monitoring. BAS collaborates with 21 other countries active in Antarctic research as part of the Antarctic Treaty, and its research covers climate change, notably the depletion of the ozone layer over the Antarctic; global warming; past climate patterns (through ice-core analysis); oceanographic processes; and geospace observations.

Scientists of the BAS were responsible for locating the hole in the ozone layer above Antarctica in 1985. They discovered the lowest concentration of ozone ever recorded on Earth. The discovery, corroborated later by other observations, was the catalyst for the

Montreal Protocol, an international agreement to phase out the production of ozone-depleting chemicals such as CFCs.

In 1992 the BAS started a major project investigating the role of the ice-covered Southern Ocean in regulating levels of CO_2 in the atmosphere. The data collected will be used to produce a mathematical model of carbon cycling at the ice edge. This research is part of an international programme.

Research by the BAS has also been done in the northern polar regions and includes collaboration with the international Greenland Ice-core Project. In 1992 a team of European scientists from eight nations, organised by the European Science Foundation and including scientists from the BAS, succeeded in drilling an ice core through the Greenland ice sheet down to bedrock 3,000 metres deep. First results show that the deepest layers of ice provide a record of climate conditions as far back as 200,000 years. The records are stored in the form of gas bubbles and dust particles contained in the snowfall from which each layer of ice was created.

Conservation

Many species of plants and animals are disappearing rapidly through over-exploitation or loss of habitat. As part of its response to the 'Earth Summit' at Rio de Janeiro in 1992, the Government has published a Biodiversity Action Plan outlining Britain's role in conserving biological diversity at home and overseas. Published in January 1994, the plan provides a strategy for the next 10 to 20 years. In it the Government commits itself to conserve and, where possible, enhance:

—the overall populations and natural ranges of native species and the quality and range of wildlife habitats and ecosystems;

—internationally important and threatened species, habitats and ecosystems;

—species, habitats and natural and managed ecosystems that are characteristic of local areas; and

—the biodiversity of natural and semi-natural habitats where this has been diminished over recent decades.

Conservation Within Natural Habitats

Considerable research and management are carried out to encourage the recovery of populations of species threatened with extinction. In 1978, for example, the otter seemed in danger of extinction in England. Of 3,200 stretches of river studied, only 1 in 16 were found to support the otter. Pesticides, pollution and hunting were blamed for the decline. By 1993, however, a study funded by the Vincent Wildlife Trust showed a very different picture: otters were found in 1 in 6 sites studied, including the River Thames. Improvement in the quality of river water, a ban on the use of pesticides such as DDT, and a ban on otter hunting have enabled otters to spread from existing populations in England, Scotland and Wales. Nature conservation agencies and the National Rivers Authority have helped the recovery by encouraging suitable management of river banks and by introducing young otters to rivers where the original populations had died out or where the existing population was so small that it was unlikely to recover without new recruits.

Conservation Outside Natural Habitats

Historically Britain has played a major role in the development of techniques for conserving species away from their natural habitats (*ex situ*). A number of institutions are internationally recognised as among the world leaders in this field, among them the Royal Botanic Gardens, Kew (see p. 45), which has been at the forefront of botanical research and plant conservation for 200 years. Funded mainly by a grant from the Ministry of Agriculture, Fisheries and Food, Kew houses the largest and most diverse collection of living plants, and the most comprehensive research collection of dried plants in the world.

Today, as natural plant habitats throughout the world are being destroyed and plants are becoming extinct, a priority of the research at Kew is to assess the value of whatever is being lost, so that a way can be found to reduce and reverse the rate of destruction of species.

Many micro-organisms, including viruses, bacteria, protozoa, and most algae and fungi are conserved in culture collections. These collections have a research and applied value, particularly in industry—for example the National Collection of Industrial and Marine Bacteria (for industrially significant bacteria). Research is carried out at British microbiological institutes to enhance preparation techniques and long-term storage methods—especially with liquid nitrogen. New protocols for preservation are being developed using cryogenic stage microscopy, enabling some species to be preserved satisfactorily for the first time.

Many zoos in Britain play a vital role in programmes for reintroducing species of international importance, such as the Arabian oryx, scimitar-horned oryx, Arabian gazelle, Mauritius kestrel and pink pigeon. Some have a wider role providing management plans for areas and species of conservation interest. London Zoo, for example, is involved in 52 nationally and internationally co-ordinated captive breeding programmes for exotic species and manages 29 such programmes.

Energy Research

Evidence linking global warming with CO_2 emissions has led to the need to make energy production, and within that power generation, less damaging to the environment.

Clean Coal Technologies

Britain has pioneered the most important advance in coal combustion this century—the technique of fluidised bed combustion. The technique allows a wide range of coals, including low quality coals, to be used and—with the addition of limestone or dolomite to the bed—leads to lower polluting sulphur emissions to the atmosphere. The British technology is employed under licence in a number of countries.

One such advanced system is the topping cycle programme, in which some of the coal is converted to hot gas for use in a gas turbine and the remaining char is burnt in a fluidised bed to raise steam for a conventional steam turbine. This offers the opportunity to reduce CO_2 emissions from coal-fired power stations by 20 per cent and is to be developed by an industry-led consortium.

Several new clean coal technology research projects have recently been announced by the Government. They are part of a range of projects which are intended to be taken forward to the pilot plant stage.

Renewable Energy

The Government supports an R & D programme into renewable forms of energy. R & D policy concentrates resources on key technologies—energy from coppice and waste, wind, solar and fuel cells—and aims to increase their appeal to investors. Planned expenditure on the Government's renewables programme for 1995–96 is £18 million.

Appendix: British Scientific Achievements

Examples of British pioneering achievements in science and technology are given in this Appendix. Britain cannot claim sole credit for all the achievements listed; some owe their success to international collaboration or to earlier developments overseas, while in others, inventions in Britain have been perfected elsewhere.

Early Scientific Achievements

Physics

Sir Isaac Newton (1642–1727) is generally recognised as one of the greatest Western scientists. His contributions to mechanics—his three laws of motion and theory of gravitation—were perhaps the most outstanding advances in knowledge made during the scientific revolution of the 16th and 17th centuries, and have been fundamental in subsequent scientific progress. He also developed one of the first forms of calculus, and, while formulating his theory of gravitation, described the motions of the planets in the solar system.

Michael Faraday (1791–1867) first demonstrated the phenomenon of electromagnetic induction, and developed the first electric motor, generator and transformer.

James Joule (1818–1889) was the first to show that heat energy and mechanical energy are equivalent.

Lord Kelvin (1824–1907) generalised Joule's work into the laws of thermodynamics and introduced the absolute temperature scale (named after him). His work on electromagnetism, together with Faraday's, gave rise to the theory of the electromagnetic field, while his work on wire-telegraphic signalling played an essential part in the laying of the first Atlantic cable.

James Clerk Maxwell (1831–1879) developed the theory of electromagnetic waves, laying the basis for later understanding of radio, radar, light and X-rays.

Mathematics

John Napier (1550–1617) invented logarithms.

Charles Babbage (1792–1871) was the first to put forward detailed proposals, in 1812, for an automatic calculating machine. His model 'analytical engine' established the basis of the digital computer.

Astronomy

Edmond Halley (1656–1742) is best known for his prediction that the comet of 1680 would return in 1758, based on his conviction that comets follow elliptical paths around the sun.

Sir William Herschel (1738–1822) discovered Uranus and the configuration of the Milky Way.

Chemistry

Robert Boyle (1627–1691) was the first to define chemical elements.

Joseph Priestley (1733–1804) discovered 'dephlogisticated air' (later named oxygen by Lavoisier) in 1774 and many other gases.

John Dalton (1766–1844) originated the theory that the atoms of different chemical elements are distinguished by their different atomic weights.

Sir Humphry Davy (1778–1829) pioneered the study of electrochemistry and also invented the miner's safety lamp.

Biology and Medicine

William Harvey (1578–1657) demonstrated the continuous circulation of the blood.

Edward Jenner (1749–1823) pioneered vaccination and established the concept of immunisation against disease when he demonstrated the effectiveness of immunisation against smallpox.

Charles Darwin (1809–1882) was the first to formulate the theory of evolution by natural selection in his book *The Origin of Species* (1859).

Lord Lister (1827–1912) was the founder of antiseptic surgery. By the use of antiseptic sprays and dressings, he revolutionised control over infection in hospitals.

Early Technological Achievements

Metals, Materials and Engineering

Abraham Darby (1677–1717) was the first to smelt iron using coke, rather than charcoal, in 1709.

Thomas Newcomen (1663–1729) built the first full-size atmospheric pressure steam engine in 1712, following (but not directly based on) Thomas Savery's working model of 1698.

James Hargreaves (1720–1778) invented the spinning jenny, a machine for spinning several threads at once.

Richard Arkwright (1732–1792) developed the first practical mechanical spinning machine using rollers.

James Watt (1736–1819) invented the condensing steam engine, a great improvement on earlier engines and which laid the foundations of modern power-driven machinery.

Edmund Cartwright (1743–1823) invented the power loom for weaving.

Sir Henry Bessemer (1813–1898) invented the Bessemer converter and process for making cast steel.

Sir William Siemens (1823–1883) invented the open-hearth furnace for making steel. (Siemens was German but worked in Britain and became naturalised.)

Sir Joseph Swan (1828–1914) invented the incandescent electric lamp (simultaneously with Thomas Edison of the United States) and also produced the first man-made fibres (achieved experimentally in the form of a cellulose-based artificial silk).

John Boyd Dunlop (1840–1920) invented the pneumatic bicycle tyre in 1888.

Sir Charles Parsons (1854–1931) invented the steam-driven turbine, patented in 1884.

Agricultural Machinery and Food Processing

Jethro Tull (1674–1741) invented a mechanical seed drill and a horse-drawn hoe.

Andrew Meikle (1719–1811) invented the first useful threshing machine in 1788.

Patrick Bell (1799–1869) built the first successful mechanised reaper in 1826.

Bryan Donkin (1768–1855) inaugurated the first food canning factory in 1812.

Vehicles and Transport

Richard Trevithick (1771–1833) built the world's first working railway locomotive, which ran for the first time at Penydarren (Mid Glamorgan) in 1804.

The Stockton and Darlington Railway, opened in 1825, was the first public passenger railway to be worked by a steam locomotive, *Locomotion*, designed by George Stephenson (1781–1848).

The world's first practicable steamboat was the *Charlotte Dundas*, launched in 1802. Used for experiments on the Forth and Clyde Canal, it was designed by William Symington (1763–1831).

The *Great Britain*, built in 1843 by Isambard Kingdom Brunel (1806–1859), was the first ocean-going iron ship and the first screw-propelled vessel to cross the Atlantic.

Civil Engineering

The first bridge built wholly in cast iron, over the River Severn at Coalbrookdale (Shropshire), was the work of Abraham Darby III (1750–1791) and was officially opened in 1781.

The first iron canal aqueduct was constructed in Shropshire by Thomas Telford (1757–1834) in 1796; he also built the Menai Suspension Bridge in north Wales, the world's largest span at the time (1826).

Joseph Aspdin (1779–1845) patented Portland cement in 1824 from which most cement and mortar are made.

Telecommunications

Sir Charles Wheatstone (1802–1875) and Sir William F. Cooke (1806–1879) invented the automatic telegraph.

Brunel's *Great Eastern* ship laid the first successful transatlantic telegraph cable in 1866.

Energy

William Murdock (1754–1839) was the first person to apply coal gas to domestic lighting when in 1792 he used gas obtained from coal to light a room in his house in Redruth (Cornwall).

James Young (1811–1883) first extracted oil from coal and shale and refined crude petroleum into different types of oil, laying the basis for the modern oil industry.

Achievements in the Twentieth Century

Some of the many examples of important scientific developments and inventions connected with Britain are outlined below. A selection of recent major research projects is given in the chapter on pp. 54–87.

Physics and Astronomy

Lord Rutherford (1871–1937) pioneered the science of nuclear physics: his research showed that radioactivity is produced by the disintegration of atoms, thus demonstrating the nuclear structure of the atom.

Sir Joseph (J.J.) Thomson (1856–1940) discovered that cathode rays were rapidly moving particles (electrons) of negative electric charge.

Sir William Bragg (1862–1942) and Sir Lawrence Bragg (1890–1971), his son, discovered the method of determining the structure of crystals by means of X-rays.

In the 1920s C.T.R. Wilson (1896–1959) invented the cloud chamber method of photographing the tracks of charged particles such as alpha particles, protons and electrons, thus making it possible to follow the movement and interaction of atoms.

The existence of neutrons was first proved by Sir James Chadwick (1891–1974), and the groundwork for quantum electrodynamics was laid by Professor Paul Dirac (1902–1984) in the 1930s.

In 1951 Sir John Cockcroft (1897–1967) was the first to split the atomic nucleus by artificial means, using an accelerator.

Sir Bernard Lovell (born 1913) pioneered the science of radio-astronomy in the 1950s (see p. 55).

In the late 1960s the science of radio-astrophysics was pioneered by Sir Martin Ryle (1918–1984) and Antony Hewish (born 1924). In 1967 their co-worker, Professor Jocelyn Bell Burnell (born 1943), discovered pulsars.

Brian Josephson's (born 1940) theoretical work in the early 1970s on superconductivity has since been used to develop higher-speed computers.

In the 1980s Professor Michael Green (born 1946) helped to develop the 'superstring' theories, which see the fundamental building blocks of matter and energy in the universe as infinitesimal 'strings' rather than points.

In 1985 scientists at Heriot-Watt University, Edinburgh, invented the transphasor, a type of optical transistor now forming the basis of a new generation of optical computers that are significantly faster than their electronic counterparts.

Much of the work on black holes (see p. 57) has been carried out by Professor Stephen Hawking (born 1942), the Cambridge astrophysicist who in 1974 showed that black holes can emit radiation with a thermal spectrum.

The first coded mask X-ray telescopes were built at Birmingham University for the United States' Spacelab 2 mission in 1985.

The world's first scanning laser microscope able to operate at some distance from its subject was built at the laboratories of the United Kingdom Atomic Energy Authority in 1984 for use with radioactive materials.

Chemistry

Dr Archer Martin (born 1910) and Professor Richard Synge (1914–1994) invented chromatography as a method of identifying and separating chemical compounds.

Professor R.G.W. Norrish (1897–1978) and Lord Porter (born 1920) developed the flash photolysis technique of analysing extremely fast chemical reactions.

Major developments in the X-ray analysis of complex molecules included the work of Max Perutz (born 1914) and Sir John Kendrew (born 1917), who explained the detailed structure of haemoglobin and myoglobin; and Dorothy Hodgkin (1910–1994), who determined the structure of penicillin and vitamin B12.

In 1985 Professor Harold Kroto (born 1939) of Sussex University discovered a new form of carbon—the C60 fullerenes. Subsequent research by British scientists has revealed the detailed structure of the carbon, which has the potential to form the basis of an ideal lubricant.

Medicine and Genetics

The hormone insulin was isolated by J.J.R. Macleod (1876–1935) in 1923 and its structure explained by Frederick Sanger (born 1918) in 1958.

In 1929 Sir Frederick Gowland Hopkins (1861–1947) was the first to show that life cannot be maintained on protein, fats and carbohydrates alone but that accessory food factors (now called vitamins) are also essential.

Interferons—proteins produced in the body to fight infections—were first isolated and named by Dr Alick Isaacs (1921–1967) and colleagues in 1957. They have been used to treat forms of cancer and can prevent transmission of some cold viruses.

In 1953 Francis Crick (born 1916), together with James Watson of the United States, made one of the most important scientific discoveries when they discovered the structure of DNA, the substance which carries the genetic code at a molecular level. Working at the MRC's laboratory in Cambridge, from the experimental results of Maurice Wilkins (born 1916) and Rosalind Franklin (1921–1958) of London University, Crick and Watson proposed the double helix structure for DNA which explains heredity in chemical terms.

The work of Sir Peter Medawar (1915–1987) in the 1950s on acquired immunological tolerance made possible the first successful kidney transplants.

Artificial hip joints were developed and first fitted by Sir John Charnley (born 1922) at Wrightington Hospital, Lancashire, in 1960.

In 1975 John Hughes (born 1942) and Hans Kosterlitz (born 1903) of Aberdeen University succeeded in isolating enkephalin, a brain chemical with some of the properties of morphine.

In 1975 Dr César Milstein (born 1927) and colleagues, working at the MRC laboratory in Cambridge, made one of the most important scientific advances of the century by producing monoclonal antibodies (antibodies produced from a single parent cell) to order. This transformed biological research and created biotechnology.

The first commercial brain and body scanner relying on computer-assisted tomography (CT) was made by EMI in 1972. Based on the technique developed by Sir Godfrey Hounsfield (born 1919), CT uses reassembled X-ray information to produce detailed cross-sections of the head and body.

The 1970s also saw the first detailed cross-section images of the body produced by magnetic resonance imaging (MRI). This technique, which involves subjecting matter to a powerful magnetic field and passing radio waves through it, can generate images of soft tissue that do not show up well on X-rays. A complete analytical instrument for MRI was developed by Oxford Instruments, working with biochemists from Oxford University who had used MRI clinically for the first time in 1981.

In 1977, when working on a virus responsible for the common cold, Dr Richard Roberts (born 1943) and the United States researcher Phillip Sharp independently discovered that genes are split up and can comprise several DNA segments.

Sir Aaron Klug's (born 1926) application of electron microscopy to the study of the structure of DNA and certain viruses in the early 1980s contributed further to understanding how cancers develop.

The technique of *in vitro* fertilisation, pioneered by Patrick Steptoe (1913–1988) and Professor Robert Edwards (born 1925), led to the birth of the world's first 'test-tube baby' in Oldham in 1978.

Professor Sir Alec Jeffreys (born 1950), working at Leicester University, developed the technique of DNA fingerprinting in 1985. This enables an individual to be identified with certainty from a small sample of tissue, blood or semen, and family relationships to be established.

Mechanical Engineering

The world's first military tank was built and used by the British War Office in 1916.

The first practicable jet engine was designed and patented by Sir Frank Whittle (born 1907) in the 1930s. Rolls-Royce built the world's first propeller-turbine (or turboprop) engine in 1945 and the world's first turbofan engine in the 1950s.

The hovercraft was invented by Sir Christopher Cockerell (born 1910) in the mid-1950s.

Joseph Bamford (born 1916) invented the world's first back-hoe loader (the 'JCB') in the 1950s, combining a hydraulically-operated excavator arm with a loader.

Electrical Engineering, Electronics and Communications

The first practical demonstration of television was given by John Logie Baird (1888–1946) in 1926; in 1938 he also first demonstrated broadcast colour television pictures.

The world's first high-pressure mercury lamp—a forerunner of the fluorescent tube—was developed by the General Electric Company in 1932, as was cold cathode and sodium lamp street lighting in the 1950s, and tungsten halogen lamps in 1961.

Long-wave radar was first used for aircraft detection by Robert Watson Watt (1892–1973) in 1935.

In 1943 Professor Max Newman (1897–1984), Donald Michie (born 1923) and Alan Turing (1912–1954) built the first electronic computer, Colossus 1, which was used for breaking enemy communications codes in the Second World War.

Professor Harold Hopkins (1918–1994) first used fibre-optics in a medical endoscope (to see inside patients' bodies) in 1954. He also invented the zoom lens in 1947.

The first laser to be used as an industrial tool was made by Ferranti in 1963.

The Post Office's first Earth station, in Cornwall, was the first to transmit live television pictures from Europe by satellite.

The world's first distributed array processor for computers was developed by ICL in the 1970s. This can carry out over 100 million calculations a second and is used to solve three-dimensional field problems in many areas of social and scientific study.

The 'Micropad', originally developed at the National Physical Laboratory and launched in 1979, was the world's first computer terminal to accept direct handwritten input.

Long-life liquid-crystal displays (LCDs), which are widely used in digital watches and pocket calculators, were first tested in the 1970s at the Royal Signals and Radar Establishment, incorporating compounds produced at Hull University.

In the 1970s BBC and Independent Broadcasting Authority engineers pioneered the development of teletext, in which pages of information are broadcast and may be viewed on television screens.

General Chemicals

Dunlop invented latex foam in 1929.

In the 1930s ICI invented 'Perspex' and discovered and subsequently produced polythene (polyethylene), possibly the world's most widely used plastic for food packaging.

Pharmaceuticals

Alexander Fleming (1881–1955) discovered penicillin in 1928; it was first produced for medical use in 1940 by Sir Ernst Chain (1906–1979) and Lord Florey (1898–1968).

Cephalosporin C, an antibiotic active against a wider range of disease-causing organisms than the naturally occurring penicillins, was first isolated by Sir Edward Abraham (born 1913) in 1953.

In 1962 the first beta-blocker successful in treating angina was introduced, having been synthesised by Sir James Black (born 1924) at ICI.

Other Achievements

Float glass, developed by the Pilkington Group in the 1950s, had replaced plate glass in the construction and transport industries by 1970. The Group has become the world's largest maker of float glass and has licensed the process worldwide.

Sialon ceramic, one of the strongest known ceramic materials, was developed in 1970 by scientists at Newcastle University. It is used as a high-speed cutting tool for machining metals.

Agricultural institutes in Cambridge were the first to achieve the birth of a calf following embryo transfer in 1966 and birth from an embryo frozen in liquid nitrogen in 1973.

Research at the Royal Aerospace Establishment at Farnborough led to the development in the 1980s of aluminium alloys containing lithium, which are stronger and lighter than existing alloys and are used in aircraft manufacture worldwide.

In the late 1970s the BP Research Centre developed a safer, cleaner and more efficient method of disposing of flare gas from offshore installations, refineries and petrochemical plants. It is estimated to have saved more than £100 million in platform construction costs and has been exported to many countries.

Abbreviations

The main abbreviations used in this book are as follows:

BAAS	British Association for the Advancement of Science
BAS	British Antarctic Survey
BBSRC	Biotechnology and Biological Sciences Research Council
BNSC	British National Space Centre
BTG	British Technology Group
CCL	Council for the Central Laboratory of the Research Councils
CERN	European Laboratory for Particle Physics
COPUS	Committee on the Public Understanding of Science
DERA	Defence Evaluation and Research Agency
DNA	deoxyribonucleic acid
DTI	Department of Trade and Industry
EMBL	European Molecular Biology Laboratory
EPSRC	Engineering and Physical Sciences Research Council
ESA	European Space Agency
ESRC	Economic and Social Research Council
EU	European Union
EUREKA	European High Technology Programme
JET	Joint European Torus
LGC	Laboratory of the Government Chemist
LHC	Large Hadron Collider
MAFF	Ministry of Agriculture, Fisheries and Food
MERLIN	Multi-Element Radio-Linked Interferometer Network
MoD	Ministry of Defence
MRC	Medical Research Council
NASA	National Aeronautics and Space Administration
NEL	National Engineering Laboratory
NERC	Natural Environment Research Council
NHS	National Health Service
NPL	National Physical Laboratory
ODA	Overseas Development Administration
ODP	Ocean Drilling Program
OECD	Organisation for Economic Co-operation and Development
OST	Office of Science and Technology

PPARC	Particle Physics and Astronomy Research Council
R & D	research and development
RTOs	Research and Technology Organisations
TCS	Teaching Company Scheme

Addresses

Government Departments

Department of the Environment, 2 Marsham Street, London SW1P 3EB.

Department of Health, Richmond House, 79 Whitehall, London SW1A 2NS.

Department of Trade and Industry, Ashdown House, 123 Victoria Street, London SW1E 6RB.

Ministry of Agriculture, Fisheries and Food, 3 Whitehall Place, London SW1A 2HH.

Ministry of Defence, Main Building, Whitehall, London SW1A 2HB.

Office of Science and Technology (Cabinet Office), Albany House, 84–86 Petty France, London SW1H 9ST.

Overseas Development Administration, 94 Victoria Street, London SW1E 5JL.

Research Councils

Biotechnology and Biological Sciences Research Council, Polaris House, North Star Avenue, Swindon, Wiltshire SN2 1UH.

Council for the Central Laboratory of the Research Councils, Chilton, Didcot, Oxfordshire OX11 0QX.

Economic and Social Research Council, Polaris House, North Star Avenue, Swindon, Wiltshire SN2 1UJ.

Engineering and Physical Sciences Research Council, Polaris House, North Star Avenue, Swindon, Wiltshire SN2 1ET.

Medical Research Council, 20 Park Crescent, London W1N 4AL.

Natural Environment Research Council, Polaris House, North Star Avenue, Swindon, Wiltshire SN2 1EU.

Particle Physics and Astronomy Research Council, Polaris House, North Star Avenue, Swindon, Wiltshire SN2 1SZ.

Others

British Council, 10 Spring Gardens, London SW1A 2BN.

British National Space Centre, Bridge Place, 88-89 Eccleston Square, London SW1V 1PT.

Royal Botanic Gardens Kew, Richmond, Surrey TW9 3AB.

Royal Society, 6 Carlton House Terrace, London SW1Y 5AG.

UK Science Park Association, Aston Science Park, Love Lane, Aston Triangle, Birmingham B7 4BJ.

Further Reading

			£
Competitiveness: Forging Ahead. Cm 2867. ISBN 0 10 128672 4.	HMSO	1995	19.50
Competitiveness: Helping Business to Win. Cm 2563. ISBN 0 10 125632 9.	HMSO	1994	15.40
Forward Look of Government-funded Science, Engineering and Technology 1995. ISBN 0 11 430131 X.	HMSO	1995	29.00
	3 volumes (including Statistical Supplement).		
Progress through Partnership: Report from the Steering Group of the Technology Foresight Programme 1995. ISBN 0 11 430130 1.	HMSO	1995	25.00
Progress through Partnership: Series of 15 Reports from the Technology Foresight Programme Sector Panels.	HMSO	1995	15.00 each
Realising Our Potential: A Strategy for Science, Engineering and Technology. Cm 2250. ISBN 0 10 122502 4.	HMSO	1993	9.65

Research and Experimental Development (R & D) Statistics 1992. Article in *Economic Trends* August 1994.
ISBN 0 11 620639 X. HMSO 1994 13.25

Each research council produces an annual report which contains detailed information about its operations.

The series of departmental reports covering the Government's expenditure plans contain information on proposed expenditure. Information about the plans of the Office of Public Service and Science is included in the report of the Cabinet Office.

Index

Printed in the United Kingdom for HMSO.
Dd.301381, 9/95, C30, 56-6734, 5673.